THE DAY
Parliament Burned Down

CAROLINE SHENTON

OXFORD
UNIVERSITY PRESS

OXFORD
UNIVERSITY PRESS

Great Clarendon Street, Oxford, OX2 6DP,
United Kingdom

Oxford University Press is a department of the University of Oxford.
It furthers the University's objective of excellence in research, scholarship,
and education by publishing worldwide. Oxford is a registered trade mark of
Oxford University Press in the UK and in certain other countries

First Edition published in 2012
First published in paperback 2013

Impression: 4

British Library Cataloguing in Publication Data

Data available

Library of Congress Cataloging in Publication Data

Data available

ISBN 978–0–19–964670–8 (Hbk.)
ISBN 978–0–19–967750–4 (Pbk.)

Printed and bound by
CPI Group (UK) Ltd, Croydon, CR0 4YY

Praise for The Day Parliament Burned Down

'A glorious micro-history... Shenton has a terrific eye for fine detail.'

Dan Jones, *Daily Telegraph*

'A hugely enjoyable read. It is formidably well researched and tells a gripping story throughout. I was riveted. Readers will be informed and enthralled by this book.'

Professor John Morrill, University of Cambridge

'One of the many achievements of Shenton's scholarly but gripping account is to revive, in all its intricacy and richness, the ghost of one of London's greatest lost treasures.'

Rosemary Hill, *The Guardian*

'No one has written about the burning of Parliament before, and this vivid, superbly researched book is a definitive account of one of the greatest cock-ups in English history.'

Jane Ridley, *Spectator*

'Could not have been bettered as our guide to this exciting event.'

Peter Lewis, *Daily Mail*

'Anyone with even a passing interest in politics or London history will be engrossed by this thoroughly researched, well-written and admirably unsensationalised book.'

David Clack, *Time Out*

'The best and most exciting and dramatic account of the burning building since Turner's paintings.'

Robert Tanitch, *Mature Times*

'Absolutely riveting... It's a thriller. Caroline Shenton is clearly one of those writers who feels that history has all the best tunes and should therefore never be boring.'

Lady Antonia Fraser

'Compelling reading... old Westminster has at last found a sympathetic obituarist.'

Times Literary Supplement

'Superb.'

Total Politics

'The detail Shenton provides is absolutely fascinating… a wonderfully detailed and gripping read.'

Daily Telegraph

'She has just the voice to narrate this tale, gripping the reader by the scruff as she describes the titanic struggle to save Westminster Hall and its stupendous hammer-beam roof. She has written a wonderful first book.'

Lucy Inglis, The Georgian

'With meticulous research, using eyewitness accounts and newspaper records, it makes for compulsive and entertaining reading.'

Sarah Clarke, Bookseller's Choice

'This is a fascinating read and I commend it to colleagues in both Houses.'

Lord Cormack, House Magazine

'Hour by hour she [Caroline Shenton] takes us through the fantastic build-up of the fire. You could have been there.'

Daily Mail

'Like a Victorian version of the TV series 24, the events are vividly counted down, hour by lurid hour, from first light at 7am to the final devastation at 6am on October 17 – and the ashes are expertly sifted for the political and social context of a calamity now mostly forgotten.'

Saga Magazine

'This excellent social history is Shenton's first book. One hopes there will be many more, not least one about today's Houses of Parliament.'

Hannah Stephenson, Liverpool Post

'Both a gripping account of that fateful night and a wide-ranging search for its ramifications across British society. Well written and extensively illustrated, this is a book that deserves attention.'

BBC History Magazine

'Makes for a truly remarkable read.'

Charlotte Heathcote, Sunday Express

'Caroline Shenton, Clerk of the Records in the parliamentary archives, shows in her excellent book, even the wood shoved into the furnaces was the product of the stranglehold of inefficient tradition.'

Jonathan Sale, Independent

ACKNOWLEDGEMENTS

Many of the illustrations in this book come from the Parliamentary Works of Art Collection at Westminster. The collection documents the history of Parliament, its buildings, people, and activities through history to the present day. For further information, and to search the Works of Art Collection, go to www.parliament.uk/art. It is full of wonderful things, including marvellously rich holdings relating to the 1834 fire and the old Palace of Westminster.

This book is a mosaic of many tiny, multi-coloured fragments gleaned from eyewitness accounts, filtered and sorted into a coherent pattern. It would have been impossible to find these effectively in the past, but today research of this kind is made much easier because of the widespread availability of online archive and library catalogues. The same goes for the commercial databases of digitized official papers, newspapers, and periodicals. I would like to pay tribute to everyone who has created and contributed to these tools which make such a difference to present-day historical research.

Philip Salmon at the History of Parliament Trust, Joe Coohill at Duquesne University, Mark Collins of the Parliamentary Estates Directorate, and Mari Takayanagi of the Parliamentary Archives read the full text of this book, and came up with numerous suggestions of various kinds. I am immensely grateful to them, but of course any errors remain my own. I would also like to thank all the librarians and archivists at the institutions mentioned in the bibliography who answered my queries about their collections, and especially those at the City of Westminster Archives Centre, Westminster Abbey Muniments, the Museum of London, the Sir John Soane Museum, Arnold Hunt at the British Library, Jane Rugg at the London Fire Brigade Museum, and William Hale at Cambridge University Library. I am in no doubt that there are other eyewitness accounts of the fire among correspondence and diaries I have not yet discovered, and I would be delighted to hear of them from interested readers.

David Adshead and Olivia Horsfall Turner generously helped me interpret floorplans of the old Palace. Members of the fire safety team at the Houses of Parliament—John Peen, Norman Davis, and Paul Kierans—explained to me how fires develop, and answered questions seemingly so stupid I was almost ashamed to ask them (such as why flames are different colours, and why smoke is black or white). Fellows of the Society of Antiquaries enthusiastically responded to my query in the SALON newsletter about the current whereabouts of Chance (sadly, I believe he may have been disposed of some time ago; his final resting place is unknown and probably unsavoury). Horatio Blood helped me locate the William Heath print of Chance reproduced in this book. Stephen Farrell kindly shared his research on the arrangement of the Armada Tapestries at the time of the fire at pre-publication stage. Thanks also to David Crook who introduced me to tallies, Henry Cole, and the Record Commission many years ago, and to Judith Flanders and Susan Shatto for their input on Dickens much more recently. The maps of 'London and Westminster, 1834' and 'The Palace of Westminster and its neighbourhood, 1834' are based on selections from the London Topographical Society's *A to Z of Regency London*, with some alterations and additions.

In Parliament I would like to thank colleagues across both Houses who have helped me: Malcolm Hay, Melanie Unwin, Emma Gormley, Susan Reynolds, and Claire Brenard of the Curator's Office; Ted Lloyd-Jukes; the Library staff of the House of Lords, especially Shorayne Fairweather, Tia Bostrom, Robert Anthony, Jo Davies, John

Greenhead, and Parthe Ward; and Elizabeth Hallam-Smith for her encouragement and enthusiasm. My colleagues in the Parliamentary Archives have been endlessly patient with my obsession with the 1834 fire and in particular I am grateful to the late Steve Chamberlain, Tim Banting, and David Trowbridge for all the copying and scanning of research sources they did for me. My good friend Julie Randles told me I needed an agent, and I found a brilliant one in Bill Hamilton: my sincere thanks to them both. My editor Matthew Cotton and all the team at OUP—Emma Barber, Jonathan Bargus, Emily Crowley-Wroe, Kate Farquhar-Thomson, Phil Henderson, Clare Hofmann, and Dorothy McCarthy—were a wonderful and enthusiastic bunch to work with. Finally, Mark Purcell has throughout provided support, ideas, inspiration, and friendly challenge in equal measure, as usual. The biggest 'thank you' goes to him.

CONTENTS

Contents

LIST OF ILLUSTRATIONS

13. 'The Destruction of the Houses of Lords and Commons by Fire on the 16th of Octor. 1834', lithograph by William Heath, published 1834. © Parliamentary Works of Art Collection, WOA 1667.
14. 'The Burning of the Houses of Parliament', watercolour by J. M. W. Turner, 1834. © Tate Britain.
15. 'The Burning of the Houses of Lords and Commons', oil on canvas by J. M. W. Turner 1834–5. © Philadelphia Museum of Art, Pa., USA.
16. 'The Burning of the Houses of Lords and Commons', oil on canvas by J. M. W. Turner, 1834–5. © Cleveland Museum of Art, Ohio, USA.
17. 'Palace of Westminster on Fire, 1834', oil on canvas by unknown artist. © Parliamentary Works of Art Collection, WOA 1978.
18. 'John Rickman, 1771–1840 Speaker's Secretary and Originator of the Census', oil on canvas by Samuel Lane, 1831. © Parliamentary Works of Art Collection, WOA 1176.
19. 'The Right Hon.ble Sir Charles Manners Sutton, G.C.B. Speaker of the House of Commons', mezzotint by Samuel Cousins after Pickersgill, published 1835. © Parliamentary Works of Art Collection, WOA 716.
20. 'The destruction by fire of the Houses of Parliament. On Thursday night Oct. 16th 1834', lithograph by M. Gauci © Mary Evans Picture Library.
21. 'Westminster Hall on Fire, 1834', watercolour by George B. Campion, 1834. © Parliamentary Works of Art Collection, WOA 1669.
22. 'Fire at Palace of Westminster from Westminster Bridge 1834', print by unknown artist, published in 1834. © Parliamentary Works of Art Collection, WOA 2572.
23. 'This View of the Remains of St. Stephen's Chapel', lithograph by A. Picken, 1834. © Parliamentary Works of Art Collection, WOA 2580.
24. 'View from Old Palace Yard after the Fire 1834', lithograph published by Vacher & Son, 1834. © Parliamentary Works of Art Collection, WOA 4648a.
25. 'House of Lords from Old Palace Yard, 1834', watercolour by R. W. Billings, 1834. © Parliamentary Works of Art Collection, WOA 1660.
26. 'View of the Ruins of the Houses of Parliament after the Fire 1834', lithograph published by Vacher & Son, 1834. © Parliamentary Works of Art Collection, WOA 4648b.
27. 'Panorama of the Ruins of the Old Palace of Westminster, 1834', oil on paper by George Scharf, 1834. © Parliamentary Works of Art Collection, WOA 3793.

LIST OF MAPS

NOTE TO THE READER

When a particular event during the fire is known to have taken place at a specific time, then that time is stated in the text. Otherwise, the many general descriptions of the fire in progress have been placed in what seemed to be the most coherent chapter, or assigned to the most likely time, as the story unfolds.

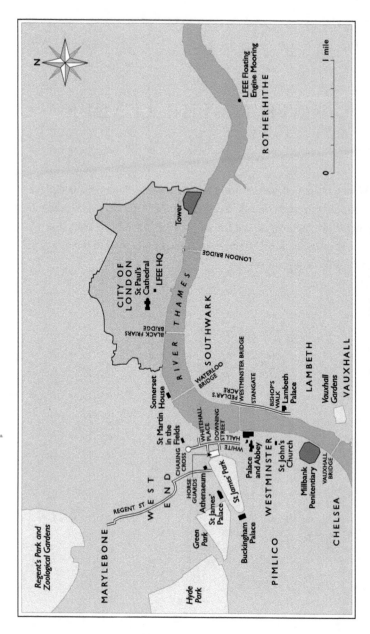

MAP 1. London and Westminster, 1834.

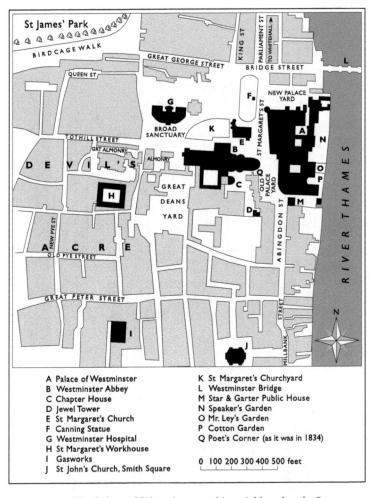

St James' Park

BIRDCAGE WALK

QUEEN ST

GREAT GEORGE STREET

KING ST

PARLIAMENT ST
TO WHITEHALL

BRIDGE STREET

NEW PALACE YARD

F.

G

BROAD SANCTUARY

K

TOTHILL STREET

GRT ALMONRY

ALMONRY

E

B

A

N

O

P

ST MARGARET'S ST

DEVIL'S

H

GREAT DEANS YARD

C

D

OLD PALACE YARD

Q

M

RIVER THAMES

L

NEW PYE ST

A C R E

OLD PYE STREET

ABINGDON ST

GREAT PETER STREET

I

J

MILLBANK STREET

N

A Palace of Westminster
B Westminster Abbey
C Chapter House
D Jewel Tower
E St Margaret's Church
F Canning Statue
G Westminster Hospital
H St Margaret's Workhouse
I Gasworks
J St John's Church, Smith Square

K St Margaret's Churchyard
L Westminster Bridge
M Star & Garter Public House
N Speaker's Garden
O Mr. Ley's Garden
P Cotton Garden
Q Poet's Corner (as it was in 1834)

0 100 200 300 400 500 feet

MAP 2. The Palace of Westminster and its neighbourhood, 1834.

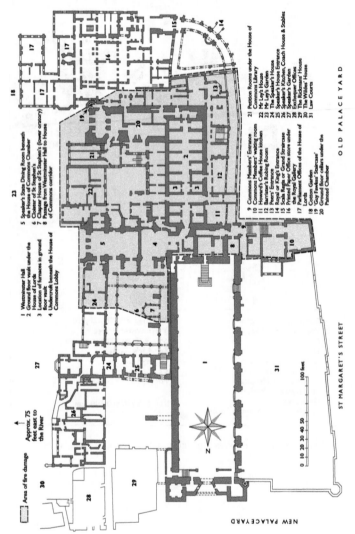

MAP 3. Ground floor, Palace of Westminster, 1834.

OLD PALACE YARD

ST MARGARET'S STREET

NEW PALACE YARD

N

0 10 20 30 40 50   100 feet

Area of fire damage

Approx. 75
feet east to
the River

1 Westminster Hall
2 Ground floor vault under the
   House of Lords
3 Location of furnaces in ground
   floor vault
4 Undercroft beneath the House of
   Commons Lobby

5 Speaker's State Dining Room beneath
   the House of Commons Chamber
6 Cloister of St Stephen's
7 Chapter House of St Stephen's (lower oratory)
8 Passage from Westminster Hall to House
   of Commons corridor

9   Commons Members' Entrance
10 Commons Members' waiting room
11 Howard's Coffee House kitchen
12 Barrister's Robing Room
13 Peers' Entrance
14 Royal Entrance
15 Scala Regia or Grand Staircase
16 Printed Paper Office store under
    the Royal Gallery
17 Parliament Offices of the House of
    Lords

18 Cotton Garden
19 'Guy Fawkes' Staircase'
20 Ground floor cellars under the
    Painted Chamber

21 Petition Rooms under the House of
    Commons Library
22 Mr Ley's House
23 Mr Ley's Garden
24 The Speaker's House
25 Speaker's House Entrance
26 Speaker's Kitchen, Coach House & Stables
27 Speaker's Garden
28 Exchequer Offices
29 The Richmans' House
30 The Winlars' House
31 Law Courts

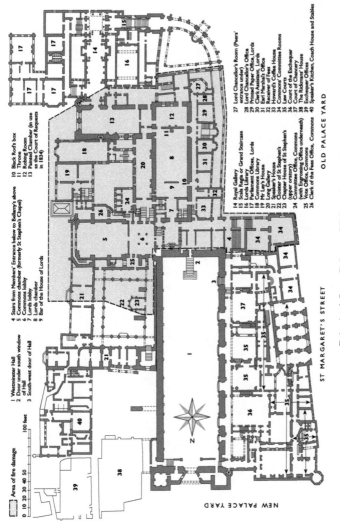

MAP 4. Principal floor, Palace of Westminster, 1834.

1 Westminster Hall
2 Door under south window of Hall
3 South-west door of Hall
4 Stairs from Members' Entrance below to Bellamy's above
5 Commons chamber (formerly St Stephen's Chapel)
6 Commons lobby
7 Lords lobby
8 Lords chamber
9 Bar of the House of Lords
10 Black Rod's box
11 Throne
12 Robing Room
13 Painted Chamber (in use as the Court of Requests in 1834)
14 Royal Gallery
15 Scala Regia or Grand Staircase
16 Lords Library
17 Parliament Offices, Lords
18 Commons Library
19 Mr Ley's House
20 Long Gallery
21 Speaker's House
22 Cloister of St Stephen's
23 Chapter House of St Stephen's (upper oratory)
24 Committee Office, Commons
25 Vote Office, Commons (with Engrossing Office underneath)
26 Clerk of the Fees Office, Commons
27 Lord Chancellor's Room (Peers' entrance under)
28 Lord Chancellor's Office
29 Printed Paper Office, Lords
30 Clerk Assistant, Lords
31 Earl Marshall's Office
32 Receiver of Fees
33 Howard's Coffee House
34 Commons' Committee Rooms
35 Law Courts
36 Court of the Exchequer
37 Court of Chancery
38 The Richmen' House
39 Exchequer Offices
40 Speaker's Kitchen, Coach House and Stables

Area of fire damage

0 10 20 30 40 50   100 feet

NEW PALACE YARD

ST MARGARET'S STREET

OLD PALACE YARD

xxi

Conflagration (n.)

A great and destructive fire; the burning up of anything; consumption by fire; severe inflammation; high fever.

Shorter Oxford English Dictionary

Prologue

EARLY IN THE evening of Thursday, 16 October 1834, a gentleman living on the New Kent Road in Southwark, on the south bank of the Thames, noticed a brilliant red light suddenly appear over the rooftops to the west. So powerful was the illumination that at first he thought a nearby house had caught fire. But then someone in the street told him a coffee-house on the other side of the river was burning. He flagged down a passing hackney carriage and, like many other Londoners that night, headed towards Westminster Bridge curious to see what was happening. The night sky was glowing with the flames, and a strong wind from the south-west blew fearsome sparks overhead. The bridge was obstructed by horse-drawn vehicles. Spectators thronged the footpaths. He could proceed no further, so he abandoned the carriage and—with an uneasy sense of foreboding—made his way to the boathouses lining the Thames at Stangate, near Bishop's Walk, Lambeth. There, turning down a narrow passage leading straight to the shoreline, he found a convenient spot which no one else seemed to have discovered. On reaching the waterside, just as he had feared, a horrifying spectacle burst upon his eye: 'St Stephen's Chapel in flames, with the House of Lords a little further to the south, and (the sensation which I felt at the sight, as an antiquary and a British subject, I shall not easily forget) the gable of Westminster Hall, contiguous to the fire, apparently alight in two or three places! The Hall (realizing the visions of the romantic age) its huge proportions, its rich-wrought and stupendous roof, were about to yield to the devouring element, and to lie a shapeless mass of ruins smouldering in the dust!'[1]

The ancient Houses of Parliament were in the grip of a tremendous fire. By dawn the following morning the great Palace of Westminster—where Parliament had sat for centuries—was a blackened, smoking, stinking ruin. Red-hot stonework, bricks, and collapsed timbers lay where they had fallen in Old Palace Yard, opposite Westminster Abbey. The molten lead which had run off the roofs in torrents was cooling in cobbled streets scattered with great piles of salvaged furniture, books, and records. Fires were still breaking out over the rooftops and around Westminster Hall, and a great plume of acrid smoke rose into the air, while on the ground exhausted firemen still worked with the army to pump the fire engines which had been in use all night. In less than twenty-four hours a terrifying event had devastated the heart of British political life, watched by thousands of spectators, and visible for many miles around London.

'Calamitous and awful', declared *The Times* leader the morning after, recognizing it as both an 'afflicting accident' and a 'spectacle of terrible beauty'.[2] For many, it struck at the essence of Britishness: 'a disastrous circumstance most deeply to be deplored by everyone who possesses a spark of patriotism and philanthropy', wrote one.[3] At first, no one knew what had caused the fire. There was speculation it might be arson, the work of a secret agent, a gas explosion, a kitchen accident, a ventilation problem, a chimney fire, or the carelessness of builders. But it was not long before people started to see something deeper and more symbolic in it. Some described it as an act of triumphant self-immolation on the part of a Parliament which had reached its highest achievement in the passing of the Great Reform Act two years previously, and which had voted through some of the most radical legislation ever passed by any Parliament.[4] Others thought it a judgement from God for that very widening of the franchise in 1832. 'Divine retribution!' Queen Adelaide is alleged to have concluded with satisfaction (a comment also subsequently attributed to the equally liberal Tsar of Russia).[5] Still more thought the firefighting efforts mirrored aspects of contemporary political indecision and bureaucratic inertia. 'In recent politics', *The Examiner* declared in a rather elaborate firefighting metaphor, 'we have had the squirt with the much ado instead of the late but potent engine. The whole calamity reads like an allegory.'[6]

It took over twenty-five years to build a new Palace of Westminster. When finished, the new building must have been profoundly shocking for those who could remember the old one. It was much bigger, and set much further back from Westminster Abbey than its predecessor. The yards in front were now vast, but the general public had less access to them. The pretty gardens leading down to the Thames on its eastern flank had been covered over with an enormous stone terrace which used every inch of space of the muddy shoreline, and even extended over the waterfront itself. The Gothic Revival detailing of the exterior convincingly echoed the decoration of the Henry VII Chapel opposite, unlike the pastiche Gothick frontage of the destroyed buildings. And inside, there was an enormous explosion of colour and design which influenced British architecture for the rest of the century and beyond. But most startling of all, the interior of the new Houses of Parliament had an ultra-regular design quite unlike the buildings it replaced. All corridors and stairs in the new Palace eventually led to its great central lobby. It was logical, monolithic, and almost obsessive in its orderliness compared with what had gone before. 'All Grecian, sir; Tudor details on a classic body', admitted its designer, Pugin.[7] For the old Palace was a glorious mess: a ramshackle, higgledy-piggledy, degraded but monumental collection of individual buildings and artworks which over the centuries had formed a conglomeration of spaces to which human beings had been obliged to adapt, rather than the other way round.

The fire was a national disaster that we have forgotten; and that disaster in turn destroyed an extraordinary complex of buildings which have faded from our consciousness. Yet before its destruction in 1834, the old Palace of Westminster had for hundreds of years been the home of monarchy, of government, of the lawcourts, and of Parliament. It was the building in which many of the great set pieces of British history took place, including the trials of William Wallace and Sir Thomas More, the Gunpowder Plot, Charles I's struggles with the Commons, Wilberforce's battle to abolish slavery, and the only assassination of a Prime Minister. And the conflagration itself was the most momentous blaze in London between the Great Fire of 1666 and the Blitz, when the new Houses of Parliament came under attack from incendiary bombs. That iconic new Palace razed to the ground the remaining ruins of much of the old Palace, and then absorbed and reworked the remainder. It also—in the process of becoming one of the most famous buildings in the world—successfully wiped our memories clean of its predecessor.

But for the hundreds of thousands who witnessed the fire on the night of 16 October 1834, it was the most memorable experience of their whole lives. It was a seminal moment in the history of the nineteenth century and Britain. Many believed they had not just watched the destruction of a national icon, but that they had also somehow experienced a turning point in the history of Parliament and of the nation. News of the disaster spread fast. Conspiracy theories were rife. A Privy Council inquiry was set up to find out the cause, while the newspapers frenziedly published stories about the historic losses, and equally, what had been saved through the efforts of onlookers from all levels of society. This, then, is the story of that fire—one which changed the face of Parliament, London, and the nation, forever.

1

Mr Hume's motion
for a new House

J OHN PHIPPS WAS Assistant Surveyor for London in the Office
of Woods and Forests. Phipps had risen spectacularly within his
department in recent years. In 1819, at the age of 23, he had
been employed as a temporary clerk on modest wages of 7s. 6d. a day,
but had quickly proved himself indispensable. He had been part of the
team responsible for designing the decorations at Westminster for
George IV's coronation festivities in 1821, and had even rolled up his
sleeves to help paint the triumphal arch for the coronation banquet in
the Great Hall—the last such feast, as it was to turn out. When a permanent job became vacant in 1822, he was the obvious candidate, becoming Second Examining and Measuring Assistant; then just seven
months later moving up to First Assistant. By 1825 he was on the next
rung, as Drawing and Measuring Clerk. Active, zealous, intelligent,
and an excellent draughtsman, he was then promoted in the major
departmental reorganization in 1832 to be Assistant Surveyor of Works
in London. Phipps's main duties were to supervise the buildings and
local managers in each London property, check requisitions, control

expenditure, and ensure good workmanship in any alterations or new constructions. This he did with great conscientiousness. Keen and energetic, Phipps had devised a new type of financial abstract to keep track of his budgets, which his superior recommended to his lazier colleagues. This cannot have endeared him much to his fellow workers, but he was known as much for his 'zeal and professional intelligence' as for his attempts to hold down expenditure.[1] The London properties he oversaw for the Office included the government buildings at Carlton House, the Charing Cross Mews, Somerset House, the Mint, Horse Guards, the British Museum, Chelsea Hospital, various prisons, plus the existing and former royal palaces of St James's, Whitehall, the Tower of London, Greenwich, Kensington—and Westminster. With such a demanding portfolio, Phipps worked long hours, which had not gone unnoticed by those above him. Three days a week in the winter he was at his desk in Whitehall Place at eight in the morning to eleven or even twelve at night. In the summer he worked Monday to Saturday from six in the morning to four in the afternoon; sometimes even on Sundays.[2]

Today, it may sound peculiar that a government department seemingly in charge of woods and forests should also be responsible for the upkeep of royal palaces. Yet in fact this is no stranger than its twentieth-century descendant, the Department of the Environment, carrying out exactly this function to 1989, at which point Historic Royal Palaces (excluding the Palace of Westminster) became a separate agency, which was subsequently turned into a charity. This strange anomaly lies in the fact that the unwieldy, but full, title of Phipps's department was the 'Office of Woods, Forests, Land Revenues, Works, and Public Buildings'. The merger, some two years before, of the old and inefficient Office of Works with the Office of Woods and Land Revenues had left some former Works staff disgruntled, disaffected, and still coping with the changes which the creation of the new office had brought about. But Phipps was not one of them. Nevertheless, he had a difficult job on his hands, and the Palace of Westminster was by far and away the most demanding—and expensive to maintain—of all his properties. In 1832 and 1833, repairs and various minor works had totalled £11,500 (approximately £33.7m in today's prices). By comparison, improvements at St James's during the same period had cost £7,900, Kensington £7,300, Horse Guards £4,300, and the Tower just £1,900.[3]

Part of the complexity was that the site at Westminster was very ancient. At this point in its course towards the sea, the river Thames runs due north before taking a sharp bend eastwards at present-day Charing Cross (*cerring* being the Anglo-Saxon word for 'bend'). The Tyburn, one of London's many lost rivers, originally flowed down from higher ground at Hampstead and split into two channels as it approached the Thames, creating a raised patch of gravel, which became known as Thorney Island. Here Edward the Confessor (1042–66) built an abbey church dedicated to St Peter to the west of the city of London, known as the west Minster, the east Minster in the city being St Paul's. Edward was immensely proud of his monastic foundation at Westminster. We know this because the embroiderers of the Bayeux Tapestry show him gazing fondly at it from his other great building project: the royal palace of Westminster. Archaeologists believe it is even possible that there was a royal palace on this site at Westminster from the decades before the Confessor. It is not unlikely that the Anglo-Danish king Cnut (d.1035) had a residence here, and it may well have been that it was on this site that he famously demonstrated to his nobles that he was not all-powerful, as he tried to hold back the tidal Thames. So in 1834 there had been a royal Palace at Westminster for nearly eight hundred years.[4]

Even the redoubtable Phipps may have been daunted by the challenge the sprawling Palace buildings presented, which by 1834 were better known as the Houses of Parliament. First of all, there was Westminster Hall, the earliest surviving building in the palace complex. Edward the Confessor's palace did not long survive the Norman Conquest. On the same site, running parallel to the river, William Rufus

(1087–1100) built his great feasting hall between 1097 and 1098, which in the centuries that followed was used by the Normans and Plantagenets as a place of ceremonial crown-wearing and the dispensing of justice. It was huge: 240 feet (73m) long and 69 feet (21m) wide, with walls measuring 40 feet (12m) high and nearly 7 feet (2m) thick. During the reign of Richard II (1377–99) the 300-year-old Hall underwent a major change. Around 1385, stone statues of all the kings from Edward the Confessor to Richard II were carved to decorate its wall niches inside—in Phipps's time, nine of them still remained in place. Then, between 1393 and 1401, one of the supreme masterpieces of European medieval art was constructed in the Hall. Hugh Herland, the King's master carpenter, and Henry Yevele, his master mason, designed and built the enormous hammer-beam roof, decorated with carved wooden angels holding shields of the royal arms, which spanned the entire width of the Hall, unsupported. It weighed an incredible 660 tons. Hundreds of prime oak trees in the woods of Hampshire, Surrey, and Hertfordshire had been felled to provide the timber for the beams, which were carved and constructed into trusses at a workshop in Farnham and then transported to Westminster on carts and by river. It was the earliest and largest hammer-beam roof in Northern Europe.[5]

In the centuries that followed, the Hall was used not only by the monarch as a ceremonial space, but also as the location for the law-courts, state trials, and political meetings; as a shopping centre, a see-and-be-seen society venue, and as a gathering place for petitioners seeking redress from Parliament.[6] By 1834, it had undergone many restorations, but parts of the walls were displaying alarming cracks. Scaffolding had therefore just been erected to begin repairs to the stonework, and to replace the flooring; major works which were being undertaken by the architect Robert Smirke (an expense which was eventually going to cost Phipps's office another £15,000).[7]

Various buildings had sprung up around the Hall in the twelfth century to cater for the royal family and the court on its visits there, the largest one being the Lesser Hall. This was built to the south of

Westminster Hall. Like the Great Hall, it ran parallel to the river, but was only half the size of its big brother. Having gone through a variety of uses over the centuries, John Phipps would have been familiar with it as the House of Lords Chamber, which it had been since 1801.[8] In 1720, importantly for our story, the Lesser Hall had been expanded upwards by the creation of a row of semi-circular 'Diocletian' windows at the top of its walls, and a new floor was therefore inserted higher up, forming a brick-vaulted undercroft at ground floor level.[9] These changes meant that visitors in 1834 would have seen nothing of this building's underlying medieval structure. The ancient window arches with their chevron decoration had been bricked up, the ceiling barrel-vaulted, and the whole was plastered over and covered with tapestries on the walls. But it was still there, lurking, beneath the surface.

Connected to the Lesser Hall, at right angles at its southern end, was another medieval building, this time built by Henry III (1216–72): the Painted Chamber. In the 1230s, the great royal connoisseur (only Charles I comes close in taste and reputation) began the modernization and complete reworking of an earlier stone chamber into a magnificent bedroom for himself. By 1236 he had commissioned the wall decorations which gave the *camera depicta* its name. Further work on the fantastic murals continued under his son, Edward I (1272–1307), so that by the fourteenth century, the name 'Painted Chamber' had stuck. This room was one of the wonders of medieval Europe. Two Irish friars on their way to Palestine in 1323 described their visit to Westminster en route, including their trip to the Abbey and 'the celebrated palace of the kings of England, in which is that famous chamber on whose walls all the warlike stories of the whole Bible are painted with wonderful skill, and explained by a complete series of texts accurately written in French to the great admiration of the beholder and with the greatest royal magnificence'.[10] When the King wished to hold a Parliament, peers and knights of the shire would travel to wherever he was on his constant itinerary around the country. So Parliament was not necessarily held at Westminster, but when it was there, it would open in

9

this great room of state, following a general proclamation in Westminster Hall. In 1834 (still with its paintings intact, some exposed through the centuries of whitewash covering them, and well-known to antiquarians) it was being used as the Court of Requests.

For his wife, Eleanor of Provence, Henry III also built the Queen's Chamber in 1237–8, to the south of the Painted Chamber and at right angles to it. The design of the Palace was now starting to look like a giant game of dominos, with each royal chamber connecting off another at ninety degrees. It was this building which Guy Fawkes had attempted to blow up (along with James I and the rest of Parliament) on 5 November 1605, by renting its ground-floor basement in which he placed thirty-six barrels of gunpowder: the Queen's Chamber being, like all principal rooms in later medieval buildings, on the first floor. The House of Lords used the Queen's Chamber for its business at Westminster from at least 1259 up to its transfer to the Lesser Hall in 1801, retiring to this room after each state opening in the Painted Chamber.[11] After the Lords moved out in 1801, the old chamber deteriorated further until finally, in the summer of 1823, it and another medieval building, the Prince's Chamber, were demolished by the eminent architect, Sir John Soane (1778–1837), who had long wished to get a commission for work at Westminster. On the razed site, Soane built a gorgeous, three-bay, neoclassical Royal Gallery, connecting the Painted Chamber to the King's Entrance and *Scala Regia*, his new ceremonial route intended for use by George IV and his successors at state openings.[12]

In the Middle Ages, the palace's main chapel was St Stephen's, first mentioned by name in the reign of King John. It ran east–west between Westminster Hall and the Lesser Hall, with its east end jutting out towards the Thames. Rebuilt by Edward I in 1292 but then abandoned, it was completed by his grandson, Edward III (1327–77), in the most magnificent style, as an English answer to the Sainte-Chapelle, in Louis IX's (1226–70) palace complex in Paris. Edward III endowed St Stephen's in 1348 with a college of canons, who were to say masses in perpetuity for him and his family. Perpetuity did not last long. When Edward's great-great-great-great-great-grandson, Edward VI, dissolved it 200 years later, he offered the redundant building to the members of the House of Commons as their permanent chamber.[13]

By the end of the seventeenth century the roof of the Commons, very high, and now 350 years old, was giving serious cause for concern. The timbers were sound but their age meant that their condition could not be guaranteed in the future. Sir Christopher Wren, then Surveyor of the King's Works, reported on its condition on three separate occasions, declaring that either the Commons needed a new home, or a radical redesign of the space was needed. The MPs chose the latter. Wren's improvements changed it significantly. In 1692, he removed the fourteenth-century clerestory and roof, and built a new roof at a lower level. He installed a new, lower ceiling below that, so that the main body of the former chapel was now divided horizontally in two. The great Gothic windows around the chapel were blocked up, with the side ones replaced by sashes and the east windows overlooking the Thames turned into three round-headed ones. The interior of the chapel was panelled over, covering up the medieval decorations and wall paintings which celebrated Edward III's devotion to his family, the saints, and the Blessed Virgin. Two galleries, north and south, were installed, supported by thin painted iron columns, with wooden capitals carved by the great woodworker Grinling Gibbons.[14] The Commons Chamber turned from a high, vast, colourful space with stone walls and exuberant decorations to a darker, more sober, wooden box, hemming in its inhabitants both from above and from the sides (and women wishing to observe proceedings were only permitted to do so through a ventilation shaft in the ceiling).

Throughout the eighteenth century the claustrophobia of the arrangements caused problems with heating and ventilation. With the impending Act of Union in 1801, matters of space became urgent. One hundred new MPs from Ireland had to be accommodated in the chamber and James Wyatt, then Surveyor of the King's Works, was asked to solve the problem. During the winter of 1800 and 1801, Wren's

wainscoting was removed, uncovering once more the medieval paintings, still in fine condition. Then, in one of the most notorious episodes
of cultural vandalism in English history, Wyatt cut out the paintings
from the walls where the new benching was to go, but not before they
had been recorded by the self-taught antiquarian John Carter, whose
vitriolic criticism of Wyatt spilled into print, where the great architect
was given a thorough roasting. Five years later, Wyatt gave a facelift to
the exterior of Wren's east windows, turning them back to a pseudo-
fourteenth-century design, in brick and stucco, so that from the river,
the window tracery at the chapel's east end looked Gothic once again.[15]
This was the Commons Chamber as it would have appeared to Phipps
in 1834.

By the early fourteenth century, the buildings within the Palace's
precincts had formed themselves into two distinct zones. The 'Privy
Palace' comprised the private apartments of the royal family, including the Painted Chamber, the Lesser Hall, the Queen's Chamber, and various other buildings long gone by the time of the fire in
1834, but including smaller chapels and further accommodation
for servants and favourites when the royal family visited. Meanwhile, the 'Great Palace', centred around Westminster Hall, included administrative, legal, and public spaces which had clustered
around that area as royal officials sought a permanent base from
which they could administer government and justice, even though
the King continued to move around the country with the court or
even go on military campaigns abroad. The Exchequer, the government's finance office and court of audit, was based in the Palace
from the early thirteenth century.[16] The lawcourts were also accommodated in or close to the Great Hall. The King increasingly
held his Parliaments at Westminster from the mid-fourteenth century onwards, and they gradually began to occupy parts of the
Privy Palace no longer in use. Over time, a series of corridors, lobbies, stairs, and connecting rooms joined the Great and Privy Palaces into a single complex. A survey of the Palace by Phipps's

colleagues, Thomas Chawner and Henry Rhodes, drawn up just two months before the fire, showed an enormously complicated arrangement of rooms which rose over three to five storeys, but with no common floor level and inserted with multiple partitions, extensions, and mezzanines, because of the random way in which its architecture had congealed together. By 1834, as well as the legislative chambers, it comprised committee rooms and offices; houses and apartments for Parliamentary officials and servants; kitchens, coffee-houses, waiting and reception rooms; two libraries; various government offices, and the lawcourts.[17]

At its northern end, outside Westminster Hall, was New Palace Yard. Meanwhile, Old Palace Yard was on the southern side of the Hall—overlooked by the flank of the House of Lords. In 1806, Wyatt had attempted to create a uniform frontage for the now chaotic arrangement of Lords' buildings, along with extra rooms and entrances. His Gothick castle design certainly created a single crenellated façade, slapped onto the buildings on the east side of Old Palace Yard, but it met with universal condemnation for its cheap and nasty execution. Wyatt's brick battlements, iron window frames, and gimcrack stuccoed oriels were met with derision. Politicians disowned it, and journalists mocked it. The architect George Dance complained of its 'beastly bad taste' and Wyatt's old antiquarian enemy, John Carter, of its 'cement influenza'. MPs were harsher. In an 1808 debate they compared its functional appearance both to a prison and to a gentleman's lavatory. To ordinary people it was simply known as the Cotton Mill. Its flimsy lath-and-plaster partition walls, covered with painted, flammable oilcloth, were essentially a piece of theatre scenery, originally intended as a temporary measure during a period of wartime economy. It turned out to be permanent and costly. The final bill for Wyatt's work (including the improvements at the House of Commons, and the Speaker's House) came to £200,000.[18] He had, in fact, created a very expensive and rickety tinder-box.

13

Housing the lawcourts was also a perennial difficulty. During the seventeenth century, the Courts of Chancery and King's Bench were housed in wooden roofless enclosures at the south end of Westminster Hall, with the Court of Commons Pleas at the north end. All were cold, draughty, and undignified spaces for legal institutions of such importance and so in the 1730s and 1740s they were all given a smarter, Gothick redesign by William Kent, later roofed over to keep out the wind.[19] This solution was still unsatisfactory, so, between 1823 and 1826, Soane had extended the eighteenth-century 'Stone Building' on the outside of Westminster Hall opposite the Abbey so the courts could have a home of their own. At the time of the fire, seven of them (the Bail Court, Exchequer, Commons Pleas, Chancery, Equity, King's Bench, and the Vice Chancellor's Court) were housed in this building to the west of the Hall: connected to the rest of the Palace, but ingeniously spliced in between its buttresses. However, due to interference by a Commons committee in Soane's original design even this new accommodation turned out to be insufficient. In 1831 a new room for the Master of the Rolls had to be created, and by the summer of 1834 there was pressure to accommodate the new Bankruptcy Court.[20]

Phipps was summoned to meet with his boss at noon on Tuesday, 14 October.[21] Alexander Milne had been appointed to the Board, as Third Commissioner, less than two months previously. Before the merger of the land side and the buildings side in 1832, he had worked in the Woods and Land Revenue Office, distinguishing himself by his expertise on royal forests and crown estates. He had even managed to ruin his eyesight through his diligent work as Secretary to the Board (a role which today we would call the Permanent Secretary of a government department); but he was nevertheless still able to undertake official duties which did not require close writing, and to have become a Commissioner had been a pleasant promotion for him, after a thirty-six-year career.[22]

Milne waved the order he had received from the Treasury dated 7 October under Phipps's nose. They were demanding that the tally office in the Exchequer buildings be cleared out; the tally foils which the Lords of the Treasury understood to be 'entirely useless' destroyed; and the empty suite of rooms there made available for the use of the new Bankruptcy Court. So the two men walked down to the rooms in question off New Palace Yard, beyond the north-east corner of Westminster Hall.[23] It was a big task: double the quantity of those wretched wooden nick-sticks needed getting rid of than on previous occasions, while another pile was to be packed away in a closet next door. There were enough to fill two horse carts. They were usually taken home for firewood by the Exchequer staff, or bundles sold as kindling, but it seemed that the latest batch had been left in this room for eight years to gather dust ever since they stopped being used. Phipps no doubt sighed: another thing to add to his long list of accommodation projects in the Palace. Milne then reminded him (as if he didn't already know) that if they were to be burned in the room's chimney, then great care should be taken, as the building was made almost entirely of lath-and-plaster and wood. Of course, Phipps knew that was a ridiculous idea. The obvious thing to do was to make a bonfire of them in the yard behind the Exchequer offices, between the buildings and the river. It would take just a couple of hours to burn them in the open air.[24]

That same afternoon, about half past three, Phipps passed on the instruction to his Clerk of Works at the Palace, Richard Weobley. Showing Weobley which sticks needed to be burnt, and which just needed storing, they both agreed that the riverside yard was a convenient and fit place to do it. However, the next morning, Wednesday, Phipps bumped into Weobley going about his business at the Palace. Overnight, the Clerk of Works had come up with an even better solution for the bonfire of the tallies. He was concerned that a fire in the open would alarm the neighbourhood, that the sticks might be interfered with, and—most objectionable of all—two cartloads of wood 'would make a great blaze and that would always collect persons'. Weobley's alternative was that they should be burnt in the furnaces which formed part of the central heating system under the House of

15

Lords. Phipps thought this was a very good plan, and said so. Fires were kept going there during Parliamentary sittings, so there was no danger in burning small quantities of tallies in the same place. He knew Weobley to be a very careful man, and that he would undoubtedly superintend the job himself.[25] Once that was settled, Phipps barely gave the task another thought: Weobley was a safe pair of hands. That evening, unable to procure a horse and cart, three labourers took a handcart, piled up their first load of tallies into it, and pushed the heavy mason's truck round to the passage leading to Cotton Garden and Mr Ley's House, a few steps inside the workmen's ground-floor entrance to the Lords. They dumped them through the archway into the vault and returned to the Exchequer, repeating this action five or six times before managing to empty the shelves. All in all, it took them about ninety minutes.[26] The workmen began burning the tallies at half past six the following morning, Thursday, 16 October, just as dawn broke.[27]

Four days later, sitting trembling before the Privy Council inquiry into the fire (a gathering which included the Prime Minister), Phipps was to burst out: 'I trust your lordships will make some allowance if I have not stated every thing, for I am in a very nervous state...I am sorry if I have done wrong...I thought there was safety there!'[28]

Members of the public were shocked when they visited the Commons in session in the early 1830s. 'I shall not soon forget the disappointment which I experienced,' said one observer,

> on the first sight of the House of Commons...it but ill accorded with the dignity of what has been termed the first assembly of gentlemen in the world...but I was not at all prepared for such a place as I then beheld. It was dark, gloomy, and badly ventilated, and so small...when an important debate occurred...the members were really to be pitied...the heat of the house rendered it in some degree a second edition of the Black Hole of Calcutta.[29]

By the early 1830s, the clamour from many MPs about the state of their accommodation had become cacophonous. Joseph Hume, the Radical MP, was their ringleader. Born in the Scottish burgh of Montrose, the son of a shipmaster, he had qualified as a naval surgeon early in life and was forever after known by his nickname, 'The Apothecary'. During his thirty-year Parliamentary career, the spectacle of the redheaded Hume, dogged and dull in his speeches, but persistent and determined in his radicalism, became a familiar sight on the Commons benches. A stubborn thorn in the flesh of any administration in power, he was one of the most regular attendees in the Chamber— often between fourteen and seventeen hours a day, fuelled by constant snacks of pears—and this had led him to become increasingly exasperated by the conditions in which he was expected to do his job.[30] Hume had been a member of a committee in August and September 1831 which had examined three well-known architects of the day (Benjamin Wyatt, Sir Jeffry Wyatville, and Robert Smirke) on how to improve the situation. The ceiling of the House was found to be too low for good ventilation and its acoustics were poor. The seats under the gallery were incredibly cramped: instead of an ideal three foot six inches in depth, the space from back to front was under two feet. The Chamber had always been unable to seat all its members, but initially there was not a difficulty. When the Commons first took over the chapel in the sixteenth century, members numbered a modest 379.[31] But their numbers had increased by nearly three-quarters over three hundred years, and the Chamber as arranged in 1834 sat a maximum of 342 MPs, and if rearranged to the ideal, would seat only 294, less than half their total number.[32] The staircase outside the Chamber leading to the Stranger's Gallery was badly positioned, and resulted in overcrowding of the Lobby by members of the public wanting to view proceedings. That was, when they could actually get into them, because often the galleries were occupied by members who could not get standing room elsewhere, while still more MPs were obliged to hang around in the coffee-house near the Lobby until the division was called, at which point they rushed out to vote. Three administrative offices near to the Chamber—the Vote office, the Clerk of the Fees' office, and the Engrossing office, were all too small for efficiency, and several more lavatories were required to keep members comfortable.[33] The report stalled at this point, possibly because the government was too preoccupied with the upheavals inside and outside the House associated with the passing of the Reform bill.

William Cobbett, the extraordinary self-taught journalist, campaigner, and MP—born a ploughboy—expressed the feelings of the radicals as they waited for an outcome:

> Why are we squeezed into so small a space that it is absolutely impossible that there should be calm and regular discussion, even from the circumstance alone? Why do we live in this hubbub? Why are we exposed to all these inconveniences? Why are 658 of us crammed into a space that allows each of us no more than a foot and a half square while at the same time, each of the servants of the King, whom we pay, has a palace to live in, and more unoccupied space in that palace than the little hold into which we are crammed to make the laws by which this great kingdom is governed?[34]

It all came to a head in two bad-tempered debates during the 1833 session. No doubt the memory of the long, tense nights spent passing the Reform bill in 1831 and 1832 had hardened opinion about the sweaty, airless, and cramped conditions of the Commons, which measured just 33 feet (10m) by 48 feet (15m) in length. There was such a lack of legroom that it was impossible for members to squeeze along the enclosed benches past the knees of their seated fellows to find an empty seat. Once they had got there, some members stayed put in order not to lose their place, and when private bills were voted on (when only the 'nays' had to leave during a division) Hume had known men sit and vote contrary to their inclinations rather than give up their seat. This was an especial problem if members had fallen asleep in the stuffy conditions. They were, in short, 'wedged in, almost like herrings in a barrel' according to Hume, when he called for another committee in March

1833. It was to take forward the recommendations of the 1831 report, putting an end to the matter, and to decide on the location and design for a new House of Commons.[35] The committee subsequently considered twenty-two plans, submitted by everyone from top architects to opinionated MPs. All manner of different shapes and styles were suggested, and the sites considered included the existing one, or extending to the east, south-east, or moving somewhere else in London completely.[36] No common theme emerged; the problem appeared insuperable to many.

The MP John Wilson Croker summed up the dilemmas. 'It would be very surprising,' he said to the committee, 'if an apartment of the most ancient palace in England, applied accidentally in the lapse of ages to a purpose for which it was not originally intended, should now be found to be, by some strange accident, the most convenient of all possible places.' And what was more, if you so altered St Stephen's as to destroy all the old associations 'you may as well go anywhere else'. It was quite maddening, and he believed

> that the accommodations about the House are notoriously imperfect, very crazy as buildings, and extremely incommodious in their local distribution…They are not well disposed for the transaction of business; they are not symmetrical with the House of Lords; they are not symmetrical with Westminster Hall; there is no proper access for Members, although we have had the misfortune to see the Prime Minister of the country murdered in the Lobby; and, on several occasions, Members have been personally insulted in going to the House. A Member who does his duty in Parliament is sometimes liable to offend individuals; he must pass every day of his life up a series of narrow, dark, tortuous passages, where any individual who wishes to insult him may have the certain and easy opportunity of doing so.[37]

The House considered the committee's report at the beginning of July. Hume introduced it. The best minds in the country, he said, had been brought to bear on the subject of improving the ventilation and seating arrangements of the place, which could only hold about half the number of members at any time. Since members had voted one million pounds for money on repairs to Windsor Castle, and an alleged £600,000 for the disastrous new-build at Buckingham Palace, Hume didn't understand why a paltry £40,000 couldn't be spent on providing for a more spacious building for the Commons. He declared that his Select Committee had decided the best situation for a new House of Commons would be due east of the current building, extending it in a direct line towards the river shoreline. The former House would become an airy lobby, with folding doors into the new, more spacious building, and this would greatly improve the efficiency of voting in the Chamber. Hume was convinced that if this plan were adopted, the Chamber might be emptied in just three minutes (rather than the current twenty), and divisions could be taken in fifteen minutes (instead of the current forty minutes it took for each member's vote to be given and registered).[38]

But opposition was strong. It was a time of economic crisis: savings were being sought from all government departments, and the budgets of the Office of Woods and Forests were a particular target. Lord Althorp, the Whig Chancellor of the Exchequer, stated that he did not believe there was anything particularly unwholesome about sitting in the Chamber (at which point another MP pointed out that Althorp himself was actually having to sit on the floor during the debate).[39] Sir Robert Peel, leader of the Tories, declared that of all the committee reports he had ever read, Hume's was the most imperfect. The report's conclusion that the current House did not provide adequate space for members was an opinion which MPs could have reached without the aid of the committee, thank you very much, and even though the committee had looked at twenty-two plans they had ridiculously failed to come to a decisive conclusion on any of them. 'Ludicrous,' shouted Peel, that all the faults of the Commons should blamed on the building. The noise, or buzz, from the considerable amount of talking in debates, and the coughing which disturbed them, were down to the behaviour of members. And to move to one of the most popular locations suggested—St James's—such a busy and bustling neighbourhood, would be very detrimental to Parliamentary business. 'Childish,' spat back Hume, of the Leader of the Opposition's intervention.[40]

So, the Honourable Members were in a state of what today might be called analysis-paralysis. To have pulled down the House of Commons Chamber and built on the same site would have been to destroy 'a great national monument, which is the pride of this country', in the words of Sir John Soane.[41] What was more, they could not bring themselves to contemplate the most radical option of all, which would have been to move Parliament and all its functions away from the riverside site in Westminster completely, and set up shop somewhere more centrally located in the burgeoning West End, either in the old Palace of St James's, or in a completely new purpose-built edifice in that quarter. Joseph Hume moved two resolutions on the night of 2 July 1833.

> First, that the present House of Commons did not afford adequate accommodation, due regard being had to the health and convenience of Members, and the dispatch of public business, and that it was necessary to erect a new building. Secondly, that an humble address be presented to His Majesty, praying that his Majesty would be graciously pleased to direct a new House of Commons to be erected, Parliament being prepared to place at his Majesty's disposal, a sum of money sufficient to defray the expense of such building.

Hume's proposals were defeated by 154 votes to 70 (a majority of 84), and the plans were shelved once again, no doubt to the fury of the would-be reformers.[42] The impasse had become notorious. It was therefore not surprising that, when the Commons Chamber burnt down just over a year later, a wag in the crowd solemnly declared, 'Mr Hume's motion for a new House is carried without a division.'[43]

2

Novelty, novelty, novelty

JOSHUA CROSS HAD been a plumber's mate with the Board of Works (as he still preferred to call it) for around fourteen years; the last three of them based at the Houses of Parliament. His father was proud to have worked there for nearly thirty-six years. Since the beginning of the month there had not been any of the usual needed on the water closets and cisterns under Mr Clerk the master plumber; but just the day before, his boss Mr Weobley had spoken to him about another task that needed doing. This was good, for someone paid a day rate. It was simple enough: some strange old sticks with writing on them needed burning and a pile of others had to be dusted and put away. He had spent the afternoon before carting the wood round to the furnaces with two other labourers, Furlong and Kirby.[1]

Nothing else is known about Kirby. But Patrick Furlong was an Irishman who had worked for one of the Board's contract paviours—Mr Johnstone—for eleven years, and before that, for another Works contractor for seven or eight years. References described him as very sober, honest, and industrious: a good, steady man. As a day labourer,

Furlong got half a crown a day in wages.[2] Only around one in twenty of the total population of London and Westminster was Irish in 1834, yet they were disproportionately some of the poorest. In 1823, a third of registered beggars in London were Irish (famine in the mid-1840s made the situation even worse). The most likely work poor London Irish would have found at the time of the fire was of the casual, labouring sort: such as paviours' plasterers and bricklayers' mates, along with all kinds of lifting and carrying jobs. This profile exactly fits what we know of Furlong. In the first half of the nineteenth century, the majority of Irish in London were located in the area between Westminster, Whitechapel, Southwark, and Holborn, with the rookery of St Giles' (now on the northern edge of London's theatreland) being particularly notorious.[3]

The Act of Union with Ireland in 1801 had abolished the Irish House of Commons and Irish House of Lords. From that point onwards, the Chambers at Westminster had to accommodate hundreds from the new ranks of members representing Ireland. It was this event that had caused both Wyatt's scandalous desecration of the Commons Chamber, and the displacement of the peers from their ancient home in the Queen's Chamber to the more spacious Lesser Hall. Irish Catholics had been eligible to vote since 1793, but until the passing of the Catholic Emancipation Act of 1829 they were unable to vote for a Catholic candidate. All the candidates were Protestant and, until the 1832 Reform Act, the constituencies were often pocket boroughs controlled by a Protestant landowner, or a Protestant corporation influenced by the local aristocracy. Would-be Catholic members of corporations were subjected to unacceptable oaths of office. But change was afoot. In 1826, the eligible electors of Waterford had chosen a pro-Catholic candidate—Villiers-Stuart—in a fight against the local Beresfords (the family of the Marquis of Waterford) who opposed Catholic emancipation. And in July 1828, Daniel O'Connell had won the Clare seat, sponsored by the Catholic Association, even though he was Catholic himself and therefore ineligible to sit in the Commons.[4] O'Connell's victory

23

highlighted how nonsensical and unjust the arrangements for voting in Ireland were. With a threat of riot and bloodshed there, the Catholic Emancipation Act, which allowed Catholics to enter Parliament, was passed rapidly, though not without considerable opposition, receiving Royal Assent in April 1829. This development proved to be the catalyst for a wider movement for reform, and inspired the work of Thomas Attwood in founding the Birmingham Political Union in December 1829.[5]

The Irish came under suspicion whenever arson was suspected. Shortly before the Swing riots in August 1830, a barn set alight at Caterham, Surrey, was said to have been caused by anti-Irish feeling, or even by Irish labourers themselves. Once the riots proper began, the press was quick to point the finger at Irish harvesters and O'Connellites: erroneously, as it turned out.[6] Catholics had also been accused of starting the fire at York Minster in 1829, before it was found to have been the work of a deranged Methodist.[7] So there were more than hints, in the days following the fire at Westminster, that the disaster was the work of an 'incendiary', and an Irish one, at that.[8] The Archdeacon of Shrewsbury felt obliged, in the public interest, to inform Lord Duncannon, the Home Secretary, that an Irish vagrant was begging in the town about a week ago, who said to a respectable inhabitant that 'there would be no good till the Houses of Lords & Commons were burnt down'.[9] He may have forgotten that Duncannon was himself an Irish peer: fourth earl of Bessborough. But by Irish, of course, the Archdeacon actually meant Catholic.

The two furnaces lay to the south of the passage which led from the Abbey side of the building down to Mr Ley's house and Cotton Garden at ground level under the House of Lords.[10] Directly above was the House of Lords Chamber. Cotton Garden, on the eastern, river, side of the Palace was named after Cotton House, a four-storey dwelling which had once sat between St Stephen's and the Painted Chamber, where the great book collector and antiquary Sir Robert Cotton had lived from 1622 to his death in 1631.[11] His magnificent collection later moved over the road to Ashburnham House in Little Dean's Yard (now part of Westminster School), where it suffered in another devastating fire in 1731, although most of the collection survived to become the foundation of what is now the British Library.

As Richard Weobley turned into the passage leading to Cotton Garden some time past seven o'clock, the first door he would have come to led to the Barristers' Robing Room, attached to Howard's coffee-house, where Counsel deposited their wigs and gowns between appearances in the House.[12] Further down the passage on the right was an ancient brick vault, locked by an iron gate: the undercroft of the Lesser Hall. On either side of the first bay as you entered the vault were the furnaces, where Cross and Furlong had been at work for half an hour or more. Each man had taken a position at the two furnaces on opposite sides of the vault.[13] The iron furnaces were about three feet high, with mouths about fifteen inches square.[14] From each a flue emerged, which ran aslant up the wall and into the ceiling void, the hot air from the fires in the two stoves regularly heating the Lords Chamber above when it was sitting, one of the flues running its sinuous way underneath the Bar of the House and Black Rod's Box, before exhausting through a chimney stack.[15] What a waste, Weobley may well have been thinking. The tallies would have made excellent firewood, and he had reluctantly turned down requests from the porter and clerk of the Exchequer Office who would both have been happy enough to have taken them home, as they had done in the past. Even cutting them up for matches had been forbidden.[16]

Weobley had told Cross the evening before that he was to burn the tallies very cautiously, putting on only a few at a time. He now watched the workmen for about ten or fifteen minutes as they cut off the strings round the bundles, damped the tallies with sprinkled water from a pail, and carefully placed half a dozen or a dozen at a time in the open stove mouths. They kept the stove doors open as the two or three layers of sticks burnt so that they could see when it was time to add the next batch. 'Cross,' he repeated, 'you must be very careful, and only put on a few at a time.' Cross passed on the instruction—again—to Furlong, busy at the other stove. There was only a very small fire going in each furnace as Weobley left to check on other works around the Palace.[17]

Weobley was probably unaware that Joshua Cross had spent three and a half years in gaol at the Millbank Penitentiary. After working for the Office of Works for some ten or twelve years as a watchman, in 1828 he thieved from a public building where he was jobbing as a plumber's mate for William Holroyd, a contractor to the Office of Works. Despite Cross maintaining that the items were 'things given to me to take away from a building; and I was stopped', the judge sentenced him to transportation to Botany Bay for a period of seven years.[18] All prisoners so convicted were sent to the Penitentiary, a modern wheel-shaped building built in 1812 less than half a mile up-river from the Palace on Millbank, where three inspectors observed the prisoners, at the end of which time the inspectors reported to the Home Secretary, and recommended the place of transportation. The number of persons in Great Britain and Ireland condemned to transportation every year in the middle of the nineteenth century amounted to about 4,000.[19] Since Cross left the Penitentiary with a testimonial of good character after three and a half years ('and five days', he pointed out), maybe he was innocent: we shall never know. However, he was re-employed on an Office of Works contract by Adam Lee, Weobley's predecessor, only weeks after his release, aided by a reference from his former master Holroyd.[20]

The 1834 fire was only the latest of a number to have destroyed and reshaped the old Palace over the centuries. In that sense, Charles Barry's new Palace of the 1840s and 1850s—so familiar today across the world from TV broadcasts, tourist brochures, and souvenir kiosks—is simply the most recent manifestation of a phenomenon that has occurred repeatedly over a period of 750 years. The first major recorded fire in the old Palace seriously damaged the Lesser Hall, the Painted

Chamber, and many other rooms in 1263. Floors and ceilings were most affected, and the murals in the Painted Chamber had to be restored or completely repainted. In 1298 there was another serious blaze in the Lesser Hall, damaging a considerable number of the royal apartments and which, this time, spread to the Abbey buildings close by, due to an unfavourable wind. A small fire even broke out in Westminster Hall during a royal banquet in 1315, but it was the 1512 disaster which had proved to be a turning point in the history of the Palace.[21] A major fire, its cause unknown, destroyed enough subsidiary accommodation around the Privy Palace to make Henry VIII and his wife of three years, Katherine of Aragon, disinclined to use it further. Portions of the crumbling walls and towers of Westminster Palace were demolished in the 1530s to provide stone for Henry's new palace at Whitehall, just a few minutes' walk away. The Painted Chamber, Lesser Hall, and Queen's Chamber seem to have been unaffected, but the royal family stopped using it as their favoured residence in the area. It may simply have been that spanking-new Whitehall was more appealing. In 1536, an Act of Parliament declared the Palace simply to be an adjunct (or 'member and parcell') of Whitehall. Westminster Hall continued to be used for coronation banquets and seasonal feasts (another fire in 1549 burnt down the palace kitchens and melted the royal silver as a result of Christmas revelries); but the Palace was now available for use—with the King's permission—to house the growing bureaucracy of government.[22]

To cope with the tidal wave of money streaming into the royal coffers as a result of the dissolution of the monasteries, four new financial offices feeding the Exchequer were established by Henry VIII between 1536 and 1542: Augmentations, First Fruits and Tenths, Wards and Liveries, and the Court of Surveyors. They were all accommodated within the Palace, and the offices located in new buildings near Westminster Hall and the existing Exchequer rooms.[23] It was the extension of the dissolution under Henry's son, Edward VI, which had brought the Commons permanently into the Palace. By secularizing colleges of

27

canons, including the one attached to St Stephen's, he was able to take the chapel into royal ownership and then offer it and its adjoining cloister for reuse by Parliament, when it met. In the sixteenth century, Parliament's usage of the Palace was by no means continuous. The Commons had moved into St Stephen's by the end of 1548, abandoning their previous peripatetic existence; their most recent home having been the refectory of Westminster Abbey. The usual pattern of sittings under Edward VI was an annual Parliamentary session, lasting around thirty to forty days. However, the Lords and Commons would not necessarily sit on every one of those days. Under Edward's sister, Elizabeth I, Parliament would meet far less frequently, with sessions many years apart. St Stephen's was therefore infrequently used during its early existence as the House of Commons.

The Chamber's physical arrangement remained much the same as when it was a chapel, influencing Commons' practice and procedure up to the present day. The rows of wooden stalls, facing one another along each side wall, greatly encouraged an adversarial debating style, and gradually became government and opposition benches; while the raised dais at the east end, where the altar had been, became occupied by the Speaker's chair so he could oversee proceedings. The distance between the two sets of benches was set by the aisle width of the chapel—sufficient for a cross-bearer and two acolytes to process to the altar before 1548. (The aisle between government and opposition was therefore not—as is often apocryphally stated—designed to keep two men brandishing swords at one another from scoring a palpable hit.) The reuse of the stone *pulpitum*—not a 'pulpit', but the thick, ornamented screen running from wall to wall across the building originally at its third bay—was particularly significant. It originally divided the chapel from its antechapel: after the dissolution the antechapel became an area where MPs could meet visitors before or after attending the Chamber, and that space eventually became known as the Commons' Lobby. Benches with their backs to the *pulpitum* at the west end of the chapel eventually became the 'cross' benches where members who

28

sided with neither the government nor the opposition sat. Access between the chapel and antechapel was through a single door in the *pulpitum* which provided a convenient method of allowing members who wished to vote against a motion to file out and be counted: it was a prototype division lobby. Finally, the *pulpitum* would have contained steps up to a gallery above for an organ or musicians to play at services. This area became a space where special visitors were invited to sit to look down on debates in the chamber. In other words, the first public gallery.[24] Meanwhile, the Lords remained happily in occupation of the old Queen's Chamber.

By 1830, the system for electing Parliamentary representatives was as complicated and maze-like as the Palace itself. The unreformed House of Commons contained 658 MPs. Of those, 465 represented boroughs, 188 counties, and five the universities of Oxford, Cambridge, and Dublin.[25] The borough—or town and city—seats in particular were a twisting rabbit-warren of different franchise entitlements. There were burgage boroughs and corporation boroughs; freeman boroughs and open boroughs; 'scot-and-lot' and 'pot walloper' boroughs: all of which required local men to possess different property-owning or residential qualifications in order to exercise their vote, depending on their location. And this decaying and complicated system, built up over centuries of accretion, was particularly prone to corruption. A number of constituencies were 'pocket' or 'nomination' boroughs: that is, the views— and therefore the votes—of the electorate in them were under the control of a single individual. In some cases, that was because he was the local landowner, to whom they owed their property or livelihood, or who had the wealth to bribe them.[26] Some were blatantly known as 'rotten' boroughs where the demography of the medieval town which formed the constituency had changed so much over the centuries that the electors were either not resident, or were now swamped by non-electors. In Gatton in Surrey, there were only six houses in the original borough, though the surrounding parish contained 135 inhabitants. In 1830 it was sold for the last time to a patron, for the enormous sum of £180,000. Another notorious example was Old Sarum, reduced since the early Middle Ages from a cathedral city to a field and an earthwork above Salisbury, but which still returned two members. Corfe Castle was just that: a ruined medieval castle. In Marlborough, Wiltshire, where the electorate was simply the town corporation, the members of the corporation were controlled by the Marquis of Ailesbury, and comprised his steward, butler, footmen, and dependants, with the mayor being his land agent. And at Dunwich in Suffolk, most of the borough had fallen into the sea; it boasted only thirty freemen voters, most of them non-resident.[27] Even Joseph Hume, the great radical himself, had entered Parliament as member for the rotten borough of Weymouth in 1812, for which he was rebuked by a contemporary.[28]

The county voting qualification was still as it had been in 1430, that is, ownership of freehold land or property worth forty shillings a year. In reality those who owned a property which entitled them to vote might also have elsewhere been tenants of a great landlord, whose influence they fell under. Where county elections were contested, huge amounts of money were expended by the candidates or the candidates' patrons to secure the vote. This was either in direct bribes or through so-called hospitality offered to the electors. In Yorkshire in 1807, a gigantic £250,000 was spent by the candidates. In Warwickshire a single candidate spent £27,000 in bribes. For many potential candidates the costs of acquiring a seat were simply too great.[29] Pitt the Younger had proposed reform in 1785, but the French Revolution and the subsequent Napoleonic wars turned the views of the governing and middle classes against anything which could open the door to the experience of the Terror of 1793–4. Anything which challenged traditional social structures and ways of doing things might encourage, wittingly or unwittingly, the overturning of everything the governing classes held dear. The effects of the French Revolution were still being felt thirty years later in England, when a peaceful protest for Parliamentary reform at St Peter's Fields in Manchester in 1819 was charged by the local militia, resulting in several fatalities and over 800 injuries, many of them to women and children.[30]

By 1830, there was intense pressure for change; but there was also intense resistance. As a result, the fight for the Reform bill was prolonged and tough, both inside and outside Parliament. It took three attempts to get it through both Houses. The first casualty of the bill was the Prime Minister, the Duke of Wellington, even before it had been introduced. He had disastrously misinterpreted the mood in the country relating to Reform. On 2 November 1830, he declared in the Lords that he was

> fully convinced that the country possessed at the present moment a legislature which answered all the good purposes of legislation...that the legislature and the system of representation possessed the full and entire confidence of the country...he was not only not prepared to bring in any measure of [Reform], but I will at once declare that as far as I am concerned, as long as I hold any station in the government of the country I shall always feel it my duty to resist such measures when proposed by others.[31]

Riots across the country followed, including a week of mob violence in London.[32] On 16 November Wellington resigned.[33] A Whig ministry took over, headed by Lord Grey. His Reform bill proposal was first presented to the Commons on 1 March 1831, and squeaked through its second reading on 23 March by 302 votes to 301. In mid-April, however, the government was defeated by a wrecking amendment designed to prevent the disenfranchisement of English voters by keeping the same number of English MPs. Parliament was dissolved and a general election called. The Whigs were returned, and the bill reintroduced into the Commons on 24 June 1831. Its second reading passed by a decisive 367 votes to 231. It emerged from its *fortieth* committee stage sitting on 7 September, and again passed its third reading in the Commons decisively. Later that month, and just before the bill entered its second reading in the House of Lords, more than 120 petitions in favour of the bill were laid on the table of the House. But the Government was defeated at its second reading by 199 to 158 votes. The bill fell. Huge popular unrest followed, and the political temperature rose to dangerous levels. A third and final attempt was then made, beginning in December 1831. It was a gargantuan constitutional effort. By the end of March 1832, a revised bill had again passed the Commons, and Grey had extracted a promise from William IV to create new pro-Reform peers if the bill looked likely to fall again in the Lords. Fulfilment of this promise was required on 9 May when a further attempt to sabotage the bill through a wrecking amendment, following its successful second reading, led the Cabinet to demand the King create fifty to sixty peerages. William refused; the Cabinet resigned; there was a run on the banks. Unable to form a government himself, Wellington advised the King to recall Grey on 15 May, and under the threat of the peerage creations (extracted from the highly distressed William IV), the Lords passed the bill by 106 to 22. The Great Reform Act received royal assent on 7 June 1832.[34]

In summary, its eventual effect in England was the following: fifty-six rotten boroughs were completely disfranchised (that is, they lost all their members); another thirty lost one of their two members; Weymouth was reduced from four to two members. They were replaced with sixty-three new borough MPs, mainly for the new industrial towns and cities in the north and midlands. The county allocations were completely revised upwards—creating an additional sixty-two MPs. Overall, England lost eighteen seats. In Scotland, there were eight new burgh seats; in Wales, three new county seats and a member each for the boroughs of Swansea and Merthyr Tydfil; in Ireland four large towns and Dublin University received five additional seats.[35] Non-resident electors in the boroughs lost the vote; and the qualification in boroughs was extended simply and clearly to £10 householders (all those occupying property with a yearly rental of this figure), but with some reservations. In the counties, new qualifications of £50 occupiers, and £10 lease- and copy-holders were created. There was a huge amount of popular support for the bill, which was viewed as a panacea, not just for the country's political ills, but also for all its social and economic ones as well. Such was the fever for reform that the wit Sydney Smith quipped at one point that every unmarried woman in the country believed the bill would get her a husband.[36]

Whether it was a landmark in the history of democracy (as it was traditionally viewed), or a damp squib, or, as the latest research has shown, it only accelerated and intensified a modest trend towards enfranchisement, is immaterial.[37] The point is, at the time, people believed it to be one of the most important political developments of their lives—for good or ill—whether they supported it or not. And some of those who opposed it felt that God was mocked through its overturning of the natural order of His creation, already threatened by the blasphemous passing of the Catholic Emancipation Act. The Duke of Wellington was accused of having called down the vengeance of heaven on the bill in debate, and the Bishop of Exeter had warned the Lords that they should vote against it and need not worry about doing so as the consequences were safely in God's hands.[38] The bill having passed, those same people no doubt felt His vengeance would not be long in coming.

Joshua Cross, plumber's mate and former inmate of the Penitentiary, lived nearby the Palace in New Pye Street.[39] This was right in the middle of a notorious area called the 'Devil's Acre' to the south-west of Westminster Abbey, centred around Old Pye Street. It was possible to take just a few steps off a respectable thoroughfare there and find oneself in the heart of one of the grimmest parts of London. For Westminster was a place of extreme contrasts: the greatest church and state buildings had grown up side by side; the crown and its subjects rubbed along together in the same buildings; some of the richest and poorest folk in the land regularly walked the same streets; and the land itself turned from solid to marshy, dry to wet, and back again, as it approached the river. And Westminster was a city not just in name because of the Abbey, but also in terms of its size: it had a population bigger than Dublin.[40]

In 1850, Nicholas Wiseman, the new Cardinal-Archbishop of Westminster, brought a new term into common English usage to describe such places. In a famous passage he described the 'labyrinths of lanes and courts, and alleys and *slum*, nests of ignorance, vice, depravity and crime' around Westminster Abbey. 'Slum' eventually replaced the older terms for houses or districts of low repute, notably 'rookery'.[41] Wiseman, born in Spain to Irish parents, trained for the priesthood in Rome and then spent a year in England in 1835, returning to the Vatican convinced that England was on the verge of a Catholic revival.[42] When he coined the term 'slum' he may well have been thinking of his earlier experiences in Westminster, as well as what he found there in 1850. The eyes of foreigners can be especially revealing. Flora Tristan, an independent and politically radical Frenchwoman (with the additional distinction of being Gauguin's grandmother), toured London in 1840. She was horrified by 'the masses of workers so thin, so pale, and whose children look so piteous', and 'the swarms of prostitutes with shameless gait and wanton glance; the bands of professional thieves; the troops of children...like birds of prey'. She had heard that of the 100,000 London girls walking the streets 20,000 died each year.[43] Disease, as one might expect, was rife across the city and the catalogue of epidemics is wearisomely shocking, even before the better-known horrors of the Victorian age. In the ten years prior to the fire there had been epidemics of smallpox (1825), typhoid (1826), rabies (1830), cholera (1832), and influenza (1831 and 1833). In addition, infantile diarrhoea was endemic and frequently fatal.[44] It is not hard to see why, given the evidence put forward by the Westminster surveyor of sewers to an inquiry some ten years later. He spoke of finding 'human beings living and sleeping in sunk rooms, with filth from overflowing cesspools exuding through and running down the walls and over the floors...I should be afraid to keep pigs in so much filth.'[45] It is likely that Patrick Furlong, too, lived in the Devil's Acre.

It was hardly surprising that the area round the Abbey was a slum. After centuries of traditional alms-giving in the area, the most desperate people were attracted there. The opportunity for begging or what was known as 'permanent mendicancy' was great, and streets around cathedrals were generally known to attract low rents as a result. During the 1833 committee hearings on improvements to the House of Commons Rigby Wason, the MP for Ipswich, struck out against the faction who believed that only a move away from Westminster to the West End would provide a suitable location for the Houses of Parliament. On the contrary, he exclaimed, if the land around Westminster were properly drained it would be the healthiest in the metropolis, since it was situated on dry gravel, whereas the higher parts of the city were clay-bound. He placed the blame for the squalid conditions of parts of Westminster squarely at the doorstep of the Dean and Chapter of the Abbey. They, he said, were owners of the Almonry 'and its disgusting neighbourhood', and the manner of short-term leasing they employed prevented investment and improvement by responsible landlords and property developers.[46] St Margaret's Workhouse nearby, and the Penitentiary on Millbank no doubt added further lustre to the slum's southern reaches, as did the Westminster gasworks. Gasworks were often set up in or on the edges of slums, as none but the most desperate would choose to work there. And it just so happened that the Westminster Gas Light and Coke company had gone into business in 1812 on Great Peter Street, south of the Abbey.[47] Flora Tristan described a horrific visit to the company's plant there, which supplied gas to light the elegant shops and Regency *palazzi* of the West End. In the furnace room, the stokers were 'joyless, silent and benumbed...expected to perform tasks beyond the limit of human endurance...naked except for scanty cotton drawers'. The men were said to die of pulmonary consumption after seven or eight years no matter how strong they were to begin with, and their only rest between shifts was to lie on sweat-soaked, coke-blackened mattresses in a cold shed outside. Flora found it a filthy place with noxious fumes, pestilential air, and infernally hot: in fact, the floor was so hot she had to jump from foot to foot as it was burning her to stand on it.[48]

During the late eighteenth century, New Palace Yard, at the north end of the Palace, outside the great door into Westminster Hall, was transformed from the outer courtyard of the medieval Palace into an open public space connected to Whitehall and the new Westminster Bridge. The gateways which enclosed the Palace precincts were also pulled down.[49] But the whole city, not just Westminster, was in a whirl of unprecedented change. The period from 1825 to 1835 saw the last gasp of many medieval institutions and buildings, long outmoded and superseded, but clinging on until someone had the energy or inclination to finally put them out of their misery: the equivalent of England's *ancien régime*, in fact. It is extraordinary to think that between 1801 and 1831 the number of inhabited houses in London increased by two-thirds and its population rose by 73 per cent to 1.65m.[50] The city was starting to expand into the modern age, yet the pillory was used as late as 1830 (and only formally abolished in 1837), while public whippings continued well into the 1830s.[51] London was an enormous building site from the mid-1820s onwards, even before the arrival of the first railway (from Greenwich to London Bridge in 1836) and the railway mania of the 1840s and 1850s.[52] Everywhere there was new development—ancient properties were being pulled down while new squares, parks, and thoroughfares were being laid out, connecting and absorbing the villages to the west and north into the growing city. The marshland of Pimlico, bordering Westminster to the south, was drained and prepared for fashionable development from the 1820s. This was done by improving the land by using imported soil from another great development downriver—St Katherine's Dock (which, incidentally, flattened the medieval hospital and collegiate foundation of St Katherine's-by-the-Tower in the process in 1828)—though building did not start in earnest until the mid-1830s.[53]

To the north, the Commissioners of Woods and Forests' plan to join the Crown farmlands at Marylebone Park to the Palace of Westminster by means of a ceremonial carriage route had originally been conceived in the 1790s. The intention was to develop the park (when the leases ran out in 1811) into a new town and pleasure ground for gentlemen who also needed to access Parliament and Whitehall for political business. John Nash's subsequent redesign of the farmland became Regent's Park. This was connected via a grand carriage drive to St James's Park, beginning at Portland Place, sweeping along the crescent-shaped temple to consumerism and high society called Regent's Street, and coming to a halt at Carlton House Terrace. Here there was easy access to the political heart of the nation: just left to the Houses of Parliament and Whitehall, or right to the King's new palace at Buckingham House. Nash's stunning piece of town planning was still new, having been completed only in 1830.[54] Strolling north up Parliament Street from the Palace in 1834 would also have revealed a scene buzzing with building activity. The Nash-inspired 'West Strand' development around St Martin-in-the-Fields was well under way. At the time of the fire, the piazza that later became known as Trafalgar Square was five years into its construction, with another seven to go; and the new National Gallery had begun to be built in 1832 on its northern side. Where the new square extended into Pall Mall, there was a new gentleman's club, the Athenaeum, founded in 1824 by John Wilson Croker, one of the MPs who had complained so vociferously about the House of Commons' accommodation.

The decade prior to the fire witnessed the most significant changes to the landscape of old London since the rebuilding projects that followed the Great Fire of 1666. The City of London corporation even demolished the famous medieval London Bridge, and replaced it with a new one slightly upriver.[55] Whitehall itself was significantly redeveloped too—Scotland Yard and the frontage of Whitehall was in a constant state of refurbishment or rebuilding for thirty-four years up to 1830, including the completion of John Soane's new Privy Council

Office and Board of Trade building in 1827. The open space today occupied by Parliament Square was not cleared of houses until the 1870s,
but one of its present-day occupants is a bronze statue of George Canning (Prime Minister for just 119 days in 1827). This monument had
been unveiled by the famous sculptor Sir Richard Westmacott, in a
prime position between New Palace Yard and the north door of the
Abbey, in 1832, and was a notable landmark at the north-west end of
the Palace. Westmacott had also been responsible for the controversial
statue of the (very) naked Achilles in Hyde Park, a memorial to the
victories of the Duke of Wellington,[56] and the sculptor played a small
part in the rescue attempts on the night of the fire. Almost everywhere
around the Palace and the Abbey was therefore new, or being renewed.
No wonder then that Ralph Bernal, MP for Rochester, considered that
by 1835, 'Novelty, novelty, novelty' was the main motivation of 'the tens
of thousands of high and low within the confines of our overgrown
metropolis'.[57] Nash's new West End was frequented by the court, the
upper aristocracy, rich artists, provincial nobility, and glamorous foreigners, in all its splendour. The houses were well built; the streets
straight and regular. The thoroughfares were filled with glittering carriages carrying magnificently dressed ladies; dandies on beautiful
horses; and gangs of liveried valets touting gold- or silver-headed canes.
The 'elegant commerce' of Regent Street and Bond Street provided
the leisured classes with extravagant new modes of dress. Nowhere was
fashion more fickle. The high-waisted dresses of the *belles* were cur-
rently being completely overshadowed by the most enormously top-
heavy wide puffed sleeves, and stiffly elaborate hairstyles held in place
by outrageously expensive headdresses. These were particularly on
show on Sundays, when an 'irresistible tide of landaus, barouches, brit-
skas, and cabriolets' sped towards the gates of Regent's Park carrying
their occupants to promenade in the new gardens of the Zoological
Society of London (designed by Decimus Burton, also the architect of
the Athenaeum, who incidentally had put forward his ideas for a new
Commons Chamber in 1833 too).[58]

People described the frequently rowdy Commons in the early 1830s as a 'bear garden'. There, the noise of debates echoed, in the words of one observer, with crowing, braying, yelping, and meowing and other 'zoological sounds'.[59] But it was not quite the same as Regent's Park. For when the royal menagerie had closed down at the very end of 1831 (the last vestige of a tradition of keeping a palace collection of animals dating back to the Middle Ages), the 150 creatures had moved from their cramped, old-fashioned quarters in the Tower of London to modern cages within the Zoological Society of London's enclosures.[60] The MPs meanwhile were still baying unsuccessfully to be let out of their ancient accommodation in the unfashionable part of town.

3

Worn-out, worm-eaten, rotten old bits of wood

'YOU'RE GOING ON very rapid, burning those sticks,' said Richard Reynolds, the House of Lords' firelighter. Cross had called him over to see the task he and Furlong were doing at some point between nine and ten, and was boasting that they had the use of the flues to burn all the wood. Standing and watching the two labourers for three or four minutes, Reynolds, whose job it was usually to tend the stoves and grates throughout the House, was irritated. The men were laying on five or six handfuls of sticks at a time. Both stove doors were open and the fuel stacks inside were more than three-quarters high, the furnace mouths stuffed with wood. Two astonishing great blazes were making a terrific noise, crackling and popping, the flames leaping up the flues, while the air inside shimmered in a heat haze. Cross later confessed that although it was true his fire was making a huge roar, he had followed Weobley's cautions to the letter: the very model of a responsible employee. There was some noise, admitted his companion Furlong, 'not a great roaring draft; but you must hear a little, of course'.[1] But how could a fire possibly have happened, they were later asked, if there was only six inches of flame, as they stated, in a furnace thirty-six inches deep?[22]

Sucking his teeth in disapproval, Reynolds wandered off down the passage to check the fireplace in the lobby where the Lords' porter usually sat. His usual job when the House was in session was to lay and light the coal fires, a task which his son David had been helping him do for the last three or four years. Those furnaces were usually filled with caked-together coal, three parts full (otherwise they would not light), which lasted for five or six hours, but 'never no accident had happened' when he had care of the flues. Sometimes the furnace doors did get red-hot (when the weather was very chilly and a big fire was needed to last all night), but the doors always remained firmly shut. Reynolds' fires were quite burnt down in the morning and it was a regular heat from the coals, not the flaming great maw which he had just seen. They used special Welsh coal for the House of Lords, which threw out a deal of heat but little fire; not ordinary sea-coal, which gave off lots of flame. But Cross and the others had a key—not his key—to the flues, so someone must know what they were about. Things hadn't been the same since the main key to the House of Lords had been taken from him and the other staff; he even used to have one for the Robing Room, but no more. Shrugging, he left for home, as he had no specific orders to be in the House.[3]

On the floor above, the housekeeper's normal routine was to inspect the Lords Chamber and check that the servants had cleaned it properly first thing. The maids usually began this work at eight in the morning, but when the House was not sitting—as now, since it was recess—they had begun at nine o'clock. As a result, the housekeeper decided to go in at ten, instead.[4] The job of Housekeeper to the House of Lords was an ancient royal appointment. In 1834 it was occupied by Frances Brandish, but in reality delegated to a deputy, called Jane Julia Wright, who lived on site at Westminster.[5] In other words, the post was a sinecure, allowing the postholder (frequently an elevated personage) to take the salary without doing a stroke of work. The deputy housekeeper

meanwhile, had all the burden of undertaking the day-to-day responsibilities of the post. She was responsible for the routine safety of the
House: except at night, or when the House was sitting. That meant
ensuring the great door from the lobby of the House of Commons remained firmly shut except when a doorkeeper was present, to prevent
strangers—particularly tradesmen and labourers—wandering through
at will from the Commons to the Lords. She was charged with ensuring
all the entrances and window shutters were closed as soon as the House
rose, and every evening she handed over bunches of keys to the Chamber, Robing Room, committee rooms, and official apartments to the
pair of watchmen who patrolled the darkened corridors, passages, and
offices at night. In the mornings, she received the keys back, supervised
the cleaning, and was available all day to show respectable visitors
around the House of Lords Chamber and the new buildings. The keys
of the committee rooms and strangers' galleries were hung up in the
Robing Room, ready for easy access by doorkeepers during daylight
hours; the other keys the deputy housekeeper would keep for herself.[6]
Jane Wright was in fact away from home on 16 October. Her mother-
in-law, Elizabeth Wright, was standing in for her, as she usually did on
those occasions.

Unlocking the west door of the Chamber at ten o'clock, Mrs Wright
observed some smoke in the body of the House. As she recalled having
seen 'a little fog or smoke' in the same place the day before, she thought
little of it. David Reynolds, the firelighter's son, was employed to watch
the west door so that strangers did not casually wander in, and had arrived on duty just before ten. Mrs Wright asked the boy about the
strong smell of burning wood. David, a bright lad, remembered seeing
the truck being used to cart the sticks round to the vault below on the
previous afternoon, and answered that it was the tallies being burnt.
Elizabeth Wright subsequently claimed that she then told him to go
and enquire about it downstairs and bring the men up to speak to her,
but according to the boy, the housekeeper simply asked him to open
one of the windows, before going back upstairs to her apartment in the

attic storey.[7] For some days afterwards, Mrs Wright remained confused (or deceived herself and others) about the amount of intervention she made that morning. David claimed only to have been sent downstairs by her to investigate in the afternoon, not before midday.[8] He certainly had noticed that there was a smell of wood smoke throughout the day, and Mrs Wright had mentioned it three or four times: but only to him, and each time he told her it was the tallies burning she seemed to be hearing the explanation for the first time.[9]

Between ten and eleven o'clock Furlong noticed Cross slip out of the vault, and remembered him being away for about a quarter of an hour.[10] The plumber's mate had, in fact, gone upstairs of his own accord, knocked at the door of the House, and asked Mrs Wright to let him into the Chamber, in order 'to see whether there was any particular heat' and to check 'how the flues was'. On letting him in, she pointed out the smell; and Cross explained that he and Furlong were burning wood in the stoves, rather than coal. According to him, she said nothing in response, and did not seem alarmed or disapproving, although he rather defensively pointed out that he was doing the work by order of Mr Weobley, so she must have queried something about it: 'I said there was no danger or something of that kind.'[11] Part of his intention, he alleged, was to check on the thermometer—or 'ventilating clock'— in the Chamber, because the flue had not been used for so long over the summer, 'and I thought there might probably be a good deal of damp air in it'. This explanation did not persuade the later inquiry about his motive for looking at it, and Cross admitted that around that time the brickwork inside his flue was not red hot, 'but looking rather red'. The thermometer was showing a cool 55°F (13°C).[12]

Like the Commons, the Lords Chamber could get stiflingly hot when full. (On the fifth night of the marathon second-reading debate of the Reform bill in October 1831, the temperature in the House had reached a suffocating 85°F (29°C). The bill fell in the Lords by a majority of 41.[13]) Cross then put his hand on the floor matting close to where his flue came up against the floor—it was not particularly hot. All of these checks, Cross claimed, were purely for casual interest: 'I had no idea of risk of fire.' Everything appeared safe.[14] Mrs Wright afterwards gave a different account of what went on during that conversation. She claimed that when Joshua Cross arrived and explained to her about the burning, she had said, 'That is very odd; who gave you leave to do it?' When Cross replied it was the 'Board of Works', she invoked the name of her immediate boss, the Yeoman Usher, saying: 'If Mr Pulman was here, he would not allow you to do it, and annoy me'—after which, Cross left.[15] These two stories do not seem to tally. A third account of Cross's visit to the Chamber came from the lad David Reynolds. From his post at the door, he saw the man checking at the thermometer, and watched as the housekeeper and Cross exchanged some words, but could not hear most of what they said. Mrs Wright had come down from her apartment specifically to show a group of visitors round, and as Cross turned to leave he said that he was doing it by Mr Weobley's order, while the housekeeper moved away in the opposite direction to attend to her tour party.[16] After Cross departed the Chamber he stopped to pass the time of day with another labourer on the way back to the vault below, and on reaching the furnaces again, told Furlong everything was all right upstairs.[17]

Mrs Wright's attention was elsewhere while she acted as tour guide. Other Thames-side royal palaces, including Whitehall, Windsor, the Tower, Richmond, Greenwich, and Hampton Court, had been tourist attractions since at least the late sixteenth century. There is no reason to believe that Westminster was any different. Since the royal family had not lived there since early that century, there would not even have been the same security concerns in letting the general public in despite Parliamentary sittings. In addition, with Westminster Abbey being one of the great sights of London and just across the road from the Palace, a tour of the two buildings was enough to occupy a morning or afternoon of any educated or curious person's time. The Abbey had issued its first guidebook by 1600. But in the royal palaces there was an alternative solution to providing visitor information. There, servants—who otherwise attended on the monarch when he or she was present—also doubled as tour guides. In fact, that is how they received most of their income from the reign of Elizabeth I onwards.[18] The deputy house-keeper was therefore entirely dependent on the fees paid to her by those she showed round the House.[19] This accounts for Mrs Wright's eagerness during the morning to attend to her visitors rather than the irritating men burning tallies during her tours. Her priority was to ensure the tourists had an enjoyable time—and would therefore tip her generously, so that she could pass this on to her daughter-in-law. It also explains why she considered the burning smell and smoke in the Chamber simply as an annoyance (and thus a threat to her guests' enjoyment), rather than as signs of an emerging catastrophe, particularly later on in the day when events took a more sinister turn. As well as the gate locks to the vaults, the locks into the Lords Chamber also seem to have been altered around that time, and only four or five people had been given the new version, one of them being the deputy housekeeper. Mrs Wright claimed to have drawn Mr Pulman's attention to the matter in the first place. She was concerned, she said, about the different persons who had got keys, perhaps imagining that by restricting access to the Chamber she was safeguarding the income of her family from those who sought to offer bootleg tours to their friends and acquaintances.[20]

Mrs Wright's actions on the day of the fire were a prime example of the risks inherent in the sinecure system, which devolved responsibility for operational matters away from the person paid to be in office. Sine-cures were common before the middle of the nineteenth century, after which public servants began to be recruited by competitive examina-tion (famously, a horror awaiting peers in the House of Lords in Gilbert and Sullivan's *Iolanthe*). Only the year before the fire, a Commons Select Committee had investigated the pay and remuneration of its administra-tive staff. Chaired by Josiah Guest MP, and including that ubiquitous reformer Joseph Hume, the committee had been quite considerably shaken by their findings. Of thirty-two clerks, two were true sinecurists and another four (the principal clerks of committees) were too elderly and infirm to undertake their duties, and so were drawing their salaries as a reward for past services in the form of a pension, while their work was delegated to deputies. Nepotism was rife. And the additional payments received by the clerks and other staff (for example, in the Serjeant at Arms' department)—namely, gratuities of various kinds, free books and stationery, ancient traditional fees from the Treasury, and attendance on committees—were enormous. Altogether the inquiry fully exposed the Commons' antiquated, extravagant, unequal, unjust, and inappropriate methods of remunerating its staff, given the changing attitudes towards modernizing the public service.[21] The system was clearly in desperate need of reform, but at the time of the fire few improvements had yet been made. An Act aimed at remedying the situation, which received Royal Assent on 13 August 1834, typified the timid and rather contradictory approach to the problem. At the next election, it stated, the Speaker's salary was to be reduced from £6,000 a year to £5,000; so long as the present Speaker was not disadvantaged if he continued in office. The salaries of the Speaker's Secretary, Clerk of the House, Clerk Assistant, Second Clerk Assistant, Serjeant at Arms, and Deputy Serjeant were all to be changed, provided nothing 'affect, alter, diminish, or take away the salary, allowance, perquisite, emolument, or House held, used or enjoyed' by the current holders of those offices. The Clerk Assistant and Second Clerk Assistant were assured of the same salary (as laid down in 1812) as their predecessors if they moved up the promotion ladder, irrespective of the Act. The sinecure offices were abolished, but the holders were to receive compensation for the inconvenience.[22] It took another sixteen years before the inquiry's recommendations were fully implemented, and a transparent system analogous with the newly reformed Civil Service put in place.[23]

One man who did the job he was paid for was Richard Weobley. He had his own key to the vault under the House of Lords Chamber, as did two of his own men, Reynolds the firelighter, Mr Bellamy's servants, and Cox, the Clerk of the Furniture. In all there might be nine people with keys allowing them access to the furnaces: but they were simply general locks for fastening the gates, not locks intended to secure them against mischief. Cross himself had a key to the gate. Weobley had given him this master key some time ago so that he could lock up the place at night and gain access to the water tanks on the roof, through the gate at the bottom of Guy Fawkes' staircase under the Painted Chamber, if he needed to.[24]

Weobley had been foreman during the building of the new Legacy Duty Office in 1818–19, and then in 1824 he was appointed Labourer-in-Trust at Somerset House, having long been in the employ of the architect Robert Smirke, who was an attached architect with the Office of Works.[25] Up to 1832, each royal palace and public building under the Office of Works enjoyed the services of a Labourer-in-Trust. Living on site, and charged with the safety of the building and its stores, he had to be, in the words of one of them: 'a practical person, a foreman of workmen, capable from his experience of making out working drawings, setting out works for the men in the building, and to be at all times present at them, to look at the correct execution and sound work, both as to the labour and the goodness of the materials'.[26]

The Office of Works had come in for strong criticism in the previous few years. It was a high-spending government department, with entrenched staff and a bizarre management structure. But there was one notorious architectural *cause célèbre*, above all, which had occupied and disgusted public opinion with its incompetence and wastefulness since the 1820s, namely the creation of a new London residence for the monarch. Buckingham Palace was simply a huge embarrassment. Between 1815 and 1824 the redevelopment of modest Buckingham House in Pimlico into a new royal residence for George IV by John Nash cost £18,900; but in the ten years after that, the works had spiralled out of control and the palace was still incomplete at the time of the Westminster fire. Its debts were running at £104,704 in 1831, with another £70,000 required to complete it. What was more, the new king, William IV, loathed his brother's last building fantasy, and was looking for

47

every excuse not to have to live in it. Riddled with defects, confused by the constant erection and then demolition of new alterations, plagued by inaccurate estimates for materials and labour, interminably delayed, and perceived as fundamentally ugly, the problems of the new Palace were firmly blamed on the Office of Works, even though Nash, as the attached architect responsible, was wholly independent of the Office and unaccountable to it.[27] Buckingham Palace was not the only Nash project which had dragged on for years, and turned into a bottomless pit of expense into which the Treasury continued to throw money. His development of Regent's Street for the Office of Works, though magnificent, had originally been estimated at £385,000. The project finally came in at £1,553,000 in 1826. In a last-ditch attempt to control government spending on public buildings, the Treasury put its foot down and severely restricted the Office of Works' budget to £25,000 in 1827 and £20,000 in 1828. The Treasury threats did no good, and by the end of the 1820s public opinion was firmly of the belief that not only did the Office of Works spend money like water, it also delighted in the ugliest designs they had ever seen. As Henry Bankes (MP for the rotten borough of Corfe Castle) put it, its projects were 'the most tasteless and the most inconveniently contrived, that it was possible to imagine'.[28]

Following an internal review, the Office of Works and Public Buildings was abolished and combined with the Office of Woods and Forests from 6 April 1832. It created a more streamlined organization, but with the cumbersome title of Office of Woods, Forests, Land Revenues, Works and Public Buildings. Lord Duncannon, the Chief Whip, previously holder of the political appointment of First Commissioner of Woods and Forests, became First Commissioner of the new Office. Benjamin Stephenson, the Surveyor General of the old Office of Works, became Third Commissioner of the new Office, and was furious to discover that the Buckingham Palace works were to be removed from his control and the whole of his former Office subsumed within the new. The merger aimed to save money. The 'injurious and costly' system of attached (what today we would call contracted-out) architects was scrapped, so that the

Office gained direct control over the design and building of its new projects. Many of the former Labourers-in-Trust were laid off, and suffered hardship as a result. The seven surviving postholders, given the more senior title of Clerk of the Works, were to personally superintend all works ordered by the Board through the appropriate surveyor, and were directed to communicate with him on all questions relating to the buildings in their charge. In addition, they were banned from doing private business on the side, or receiving any gifts or tips, which had led to corruption.[29] The new structure was far more efficient and less corrupt, but not achieved without a good deal of unsettling change and personal friction.

Weobley's appointment in 1832 as Clerk of Works at the Palace of Westminster was therefore a direct result of the reshuffle of those posts. His predecessor, Adam Lee, the former Labourer-in-Trust, was by all accounts a difficult subordinate: insolent, disobedient, and independent. Lee moved to become Clerk of Works at Whitehall and Horse Guards in the 1832 reorganization, but evidently found it hard to let go of his previous responsibilities, which he had held since 1806, as he had the temerity to lay his own (rather unimaginative) plans for the rebuilding of the House of Commons before the 1833 Select Committee on accommodation.[30] In replacing him with Weobley, the Office may have believed it would have a more willing and obedient manager at Westminster. With a government reshuffle in July 1834, Lord Duncannon became Home Secretary, but still kept a watching brief over the Office because of his involvement with Buckingham Palace. His replacement as First Commissioner was John Cam Hobhouse, Lord Broughton.[31] It is within this context, then—nearly three years of organizational and political upheaval surrounding the management of royal palaces—that the 1834 fire took place.

At around eleven o'clock, Weobley returned to check on progress at the flues. He asked Cross, who was much the older of the two men and whom he treated as the foreman, if the work was going all right. He was assured it was going well. To Weobley, the fires appeared to be much the same size as they had been at half past seven—that is, small. They weren't blazing, he was certain, he told the Privy Council afterwards. The furnace doors were open, and the wood inside never reached more than about four inches high while he was observing. Certainly coal fires could often be larger, and the door was not red hot, since it was never closed. Weobley was content and left the men to it, after having given another warning to be careful and to continue to put the tallies in gradually.[32]

Tallies are the most ancient form of accounting used in England. In fact, 'The Tallies' was the original name for the Exchequer before the twelfth century. By 1118, the Exchequer was the administrative offshoot of the royal court receiving and issuing payments to and from the Crown, as well as a court of audit which called before it those who owed money to the King and, over the course of time therefore, the government. It was the chequered cloth or *scaccarium* used like an abacus to calculate sums owed on the counting table which gave the Exchequer its name, but the use of tallies pre-dated even that.[33] Although the Exchequer began to wither away in the seventeenth and eighteenth centuries (while one of its own sub-offices, the Treasury, flourished and eventually took over the main business of holding the government's purse-strings), the chief financial minister in the UK government is still known as the Chancellor of the Exchequer—rather than the Treasurer.

The administrative practice of the Exchequer was first written down in 1179 in a treatise called the *Dialogus de Scaccario*, or 'Dialogue of the Exchequer', by Richard FitzNigel (1130–98), one of the king's clerks and a future bishop of London.[34] A tally was a notched wooden stick used as a receipt for government income in the Middle Ages, and beyond.[35] The word 'tally' comes from *talea*, the Latin word for a slip of wood, and not—as many people assume when they hear it—directly from the French *tailler*, meaning 'to cut' (a word which itself is derived from the word *talea*). They were an ideal form of record: lightweight, compact (easily stored in a pocket or bag), and virtually impossible to forge or duplicate. Tallies were roughly as long as the span of an index finger and thumb, and were made of hazel, box, or willow, the bark shaved off and the wood shaped into a stick with four straight sides by a special official in the Exchequer of Receipt called the tally-cutter. (The Receipt was that portion of the office devoted to physically handling income and expenditure.) The tally was then ready for use. Notches were cut into the raw wood, the size and quantity of which indicated a sum of money received by the Crown. A notch the thickness of the palm of the hand meant £1,000; the width of a thumb meant £100; £20 was the width of a little finger; and one pound was the size of swelling barleycorn. A shilling was the size of two cuts, thus making a small notch, and a penny was a single notch which did not remove any wood. A halfpenny was indicated by a small circular punch in the wood. Where a large sum of money was concerned, the thousands and hundreds of pounds would be cut on one face of the stick; the pounds, shillings, and pence on the opposite. Once the carving was complete, the tally was then cracked open lengthways by the tally-cutter with a blade and mallet, the split parting each notch on the stick vertically into two.[36] The two halves were of uneven size, as the tally-cutter would stop slicing about one inch from the end of the tally, and would then cut sideways, producing one piece which was a chunky 'L' shape—known as the 'stock' (or 'counterfoil')—and one shorter straight piece called the 'foil'. The tally-writer then inscribed the remaining smooth, uncut sides of the two parts with details of the person to whom the tally was going to be given, and the reason for the payment for the sum.

Sheriffs and other royal officials collecting taxes and fines in the counties would present themselves at the Exchequer twice a year to pay in the money they had accumulated which they owed the King's government. Each year, at Michaelmas, 29 September, they would leave the Exchequer offices quit of the previous year's account, except for any arrears owing, but with a new sum to collect for that financial year. At Easter they would return to pay in a six-monthly instalment and a tally would be cut as proof of receipt of that sum. The sheriff would leave with the stock, and the foil would be placed in the relevant county 'forel', or storage container. It was then retrieved the following Michaelmas when the sheriff paid in that year's second and final instalment, the Exchequer official reuniting the foil and counterfoil, checking they matched perfectly to ensure the sheriff had not tried to tamper with the amount which signified what had already been received. A preliminary account was drawn up based on payments during the year, and the complete tally was retained by the Exchequer to indicate that the sheriff was quit of his debt, providing there was no sum outstanding. So long as the Exchequer only had half of the tally-stick for a sum collected, the sheriff was still liable for the remaining instalment of that year.

Tallies and the accounting practices around them gave rise to a number of terms which remain in use today. The phrase 'tallying up', to mean matching two items or thoughts together, or a final accounting, is still in common English usage. 'Keeping tally' is also a familiar phrase, sometimes involving the five-bar-gate notation method (itself a visual representation of notching). More surprising financial derivations from the tally process are the words 'stock' and 'stockholder', as from no later than the 1290s tallies were in use as credit notes, which could be transferred to third parties in lieu of another debt. And the Latin word *chacia* or *scacchia*, originally another name for the stock, turned into 'cheque'. 'Counterfoil' is still familiar to many as the term for a cheque or ticket stub, or other proof of a transaction kept by the issuer of a money payment.

Amazingly, this system was still routine government accounting practice until the last quarter of the eighteenth century. The use of tallies was then formally abolished by statute in 1783.[37] But in a striking example of how the sinecure system acted as a brake on the wheels of change, the officials of the tally office were permitted to remain in post for life after 1783; and so long as they did, the ancient tally-cutting process continued.[38] After all, it provided the two most senior—the

Chamberlains—with a nice little earner in the form of £59 a year, for which they did absolutely nothing. Only on the death of the last of these two men, Montagu Burgoyne (appointed in 1772) and Frederick North, fifth earl of Guilford (appointed in 1779), on 10 October 1826 did tallies cease to be cut and the medieval procedure at last came to an end.[39]

In 1834, the Exchequer of Receipt itself was finally closed down, just six months before the fire.[40] Because the Exchequer was very ancient and had played such an important role in the administration of medieval and early modern England, its buildings at Westminster were both prominent and fixed. By 1244, the Court of the Exchequer (sometimes known as the Great or Upper Exchequer) was situated in a building outside the buttresses at the north-west corner of Westminster Hall. Extended and rebuilt in the sixteenth century, it remained on that site until 1823. The administrative offices of the Exchequer of Receipt (the Lower Exchequer) were originally situated to the east of the Hall; by 1834, they were contained within a range of red-roofed buildings on the eastern side of New Palace Yard, running parallel to the river.[41] It was from this latter building that Cross, Furlong, and Kirby had taken the tallies on the evening on 15 October. The 1834 fire, then, was both a literal and metaphorical bonfire of the sinecures: and of the state's old way of doing things.

As early as 1789, a group of fourteen of the country's top architects, including Adam, Soane, Cockerell, and Wyatt, reported to the House of Commons on their worries about the fire hazard which the rambling Palace by then presented. This led to the removal of numerous temporary wooden buildings, and ultimately the construction of new lawcourts, including the Court of the Exchequer, in the 1820s.[42] It was claimed that when a new heating system went into the Commons in 1822, using the fashionable French Chabannes method, the installers had fears it might not work, or else be dangerous. It was alleged that one engineer had said to another: 'Well, if they are dangerous, never mind, for Mr —— (mentioning the name of an eminent architect), I have often heard say, would give £500 or £1000 to see the House of Commons burnt down or destroyed...it has for years been the grand object of Mr ——'s ambition to have an opportunity of submitting a plan for the erection of such a building for the Commons' House of Parliament to assemble in.' This rumour circulated at the inquiry after the 1834 fire; and even though the shorthand writer was ordered to leave the name of the eminent architect out, it would have been all too obvious to contemporaries who this slander was aimed at.[43]

In 1822, John Soane had finally received a commission to build new lawcourts at Westminster, after many years of being thwarted in executing designs for the Palace despite being one of the attached architects to the Office of Works. The festivities associated with George IV's coronation the previous summer had required William Kent's Gothick partitions enclosing the courts of King's Bench and Chancery at the south end of the Hall to be cleared away. Without the courthouses' 'farrago of pinnacles and pineapples' (in the words of John Carter), the Hall emerged in all its magnificence at the coronation banquet, to the admiration of everyone. The Prime Minister accordingly demanded that a better place be found to house the courts. Soane was faced with the challenge of somehow squeezing the new courts into the spaces between the buttresses on the outside of Westminster Hall, opposite the Abbey. There was already a block there, dating from the previous century, known as the 'Stone Building', with a Palladian frontage, which Soane extended and wrapped around the rectangular space to form the new Law Courts.

As with so many of the nineteenth-century alterations to the old Palace, the project became mired in controversy. Soane was undoubtedly a genius, but, like most geniuses, an uncompromising one. He could not bear to build in Wyattesque crenellated Gothick, so his entire frontage was classical, to match the existing Stone Building. The design then had to be altered when a Commons committee objected to the view—from New Palace Yard—of the north end in Soane's innovative, daring style, when juxtaposed with the medieval entrance into Westminster Hall. Soane was furious and humiliated by what he called the 'clumsy, puerile and disgraceful' Gothic north end he had to slap on to the Palladian building to placate the MPs. All this seems strangely familiar: arguments between innovative architects and those who view themselves as guardians of good taste are nothing new. The Law Courts at the front of Westminster Hall were still in place at the time of the fire in 1834, and remained there until a new complex was built in the Strand in 1883, when they were finally separated from the Houses of Parliament after nearly eight centuries.[44]

During the course of the Law Courts work, Soane had to demolish the Upper Exchequer's sixteenth-century buildings, and the records in there were removed. So too were the records of the King's Bench and the Chancery previously located in Kent's pineapples. In early 1822, a large temporary shed was built to hold them in Westminster Hall prior to the demolition.[45] For the rest of the 1820s they stayed there. Today, historians may well shudder at the thought of some of the most important medieval public records being kept in this exposed location, but worse was to come. Around 1831, the first job the labourer Cross had got when he returned to the Office of Works after his stint in the Penitentiary was to help in the removal of those records to the Royal Mews in Charing Cross, now an empty space since the transfer of the horses and carriages to Pimlico in the mid-1820s.[46]

There, thousands of miscellaneous Exchequer records were heaped together in two giant constructions rather like modern Portakabins: 4,136 cubic feet (117 cubic metres) of unique historic parchments and papers dating from the twelfth to the nineteenth centuries. By 1835 they were damp, decayed, gnawed by vermin, and stuck to the stone walls of the Mews. When the archivist Henry Cole, of the Record Commission, began work on them in 1833, the appalling job of undertaking an initial sort of the records had to be performed in the most disgusting circumstances. Cole recounted afterwards how three Irish labourers, and a rat-catching dog, spent a whole fortnight in removing the Exchequer records from the sheds and placing them in sacks. As they worked, a dead cat was found in the records, as well as six or seven whole rat skeletons embedded in the parchments, not to mention the scattered bones of many others who had failed to make their way out of this archival hellhole. The dog chased the live rats from their nests as the labourers worked. 'Nothing but strong stimulants sustained the men in working amongst such a mass of putrid filth, stench, dirt and decomposition', Cole later recalled. Four thousand bushels of records finally filled 500 sacks, separated from 24 bushels of dirt, dust, and debris.[47]

The parlous state of the public records at the time merely echoed the terrible degradation of the Palace of Westminster. The late Georgian period did have its antiquarian champions, but they remained outside mainstream opinion, which cared little for the fate of national monuments and muniments, regarding such campaigners as eccentric or worse. It took the 1834 fire to jolt public conscience out of its complacency. With his intimate knowledge of the buildings, John Soane knew a catastrophe was only a matter of time. In his 1828 prospectus, *Designs for Public and Private Buildings*, he wrote of the Lords' buildings: 'the want of security from fire, the narrow, gloomy and unhealthy passages, and the insufficiency of the accommodations in this building are important objections which call loudly for revision and speedy amendment'. Of course, part of this urging was to convince the Parliamentary authorities to commission him to replace the old buildings with newer, and therefore safer, designs by his own hand; not to mention his loathing of James Wyatt. Nevertheless, it can have given him little satisfaction, six years afterwards, to see his uncanny prophecy about the Cotton Mill come true:

> In the year 1800, the Court of Requests was made into a House of Lords; and the old buildings of a slight character, several stories in height, surrounding the substantial structure, were converted into accommodations for the officers of the House of Lords, and for the

necessary communications. The exterior of these old buildings, forming the front of the House of Lords, as well as the interior, is constructed chiefly with timber covered with plaster. In such an extensive assemblage of combustible materials, should a fire happen, what would become of the Painted Chamber, the House of Commons & Westminster Hall? Where would the progress of the fire be arrested?[48]

Where, indeed?

Thursday, 16 October 1834   3 p.m.

Manifest indications
of danger

TALLY-BURNING CONTINUED THROUGHOUT the day. High water at the new London Bridge was reached at twenty past one, and then the Thames began to recede, gradually revealing its muddy shoreline upriver at Westminster.[1] Mrs Wright does not seem to have been much bothered about the increasingly peculiar condition of the Chamber until around three o'clock. On going into the Lords once more, she sent for the foreman in charge of the matting. When he arrived, she told him that she did not like the appearance of the House, it being very hot below the Bar. John Jukes ripped up the 'bump'—the painted canvas floor covering—from off the stone flags by Black Rod's Box. The Box was an enclosed wooden bench to the right of the Bar of the House, which was a barrier at the north end of the room, where strangers having business in the Chamber would stand to address the House. Feeling the paving, Mrs Wright was instantly obliged to take her hand away because of its scorching heat, and Jukes found it was hot under his feet; much too hot for him to touch with a bare hand. The bump, he said, was 'quite in a sweat' from condensation. He also noticed some smoke in the air which had not been there before.[2] The boy David Reynolds was accordingly told to open another window. At 60°F (16°C), the ventilating clock was five degrees higher than it was at eleven o'clock.[3] Today, this would seem a cool room, but in 1834 it was obviously cause for concern, not least because the upward creep in temperature was coming from the intense heat radiating from one small corner of the Chamber.

Jukes was of the opinion that the situation was rather dangerous. Mrs Wright agreed, and claimed she had sent to Mr Weobley's men several times during the day and that 'they had not attended to it and that she would write off to the office about it'.[4] She told David Reynolds to go down 'and tell Cross she was afraid there was something wrong; that the heat of the House was such that she could not stand it; likewise she could not stand on the bump at the Bar'.[5] For the first time that day, one of the Palace servants went downstairs to see what was going on. David announced to Cross that 'the House is in a complete smother, and the throne can scarcely be seen'.[6] The boy added that he thought the flues were too hot—or at least, not very cold. He fiddled curiously with a few tallies for a while then left, after asking Cross how long it would be before they would be done.[7] Mrs Wright's hope that the boy could, or would, force the men to stop was in vain. Reynolds junior came back with the message from Cross that the housekeeper had no occasion to be alarmed; he would make sure that everything was safe; the smoke couldn't be helped; and that he and Furlong would be finished within the hour, as there were not many more to burn. Mrs Wright later expressed her vexation that despite her message 'they would go on with it', and Jukes noticed her dissatisfaction with Cross's answer at the time. But she made no further remark, nor took any other action.[8] For his part, Cross had seen no occasion to return upstairs to check all was well after his single visit mid-morning.[9]

Shortly after this exchange, towards four o'clock, the Clerk of Works returned to the furnaces for his third and final check of the day. To Weobley, the situation appeared exactly the same as on his previous visits. Small fires were burning in the furnaces, sticks were not being forced up the flues, and the labourers were following his instructions closely. He claimed later to the Privy Council that when he put his hand on one of the furnaces it was as cool as usual. There were piles of ash on the brick floor, which had been drawn out seven or eight times during the day. Just a few more sticks were left to burn. Weobley was keen to check the men did not sneak off with the last sticks.[10] His mind was running on the risk of pilfering (or 'embezzlement') of firewood, but not on safety. That was the only anxiety he had in relation to the work.[11] He seemed to have entertained no suspicions that, between his visits, the pattern of stoking the stoves could have been very different. He could not see what motive the men might have had to disobey his orders while he was occupied elsewhere.[12] Weobley, by accident or design, was also completely ignorant of any of the feeble concerns being expressed upstairs. He never observed any smoke in Cotton Garden passage, did not hear of Mrs Wright's alleged complaints, and Cross and Furlong said nothing of the visits by both Reynolds, father and son.[13] It never occurred to him to go into the House during the day, as it had never occurred to him there was any danger. If it had, he later testified, he would never have arranged the burning of the tallies where he did.[14] Soon after Weobley departed, Cross and Furlong placed the final sticks in the stoves, thus finishing their hot and tedious task. They left the vault shortly after four o'clock and went off to have some well-earned beer at the Star and Garter public house in Abingdon Street, directly opposite Soane's King's Entrance to the Lords. The men did not report back to Weobley when they knocked off. Joshua Cross found it difficult to account for the later fire, given the care which he claimed he and his companion had applied to their work.[15] 'It never ran in my head there was any danger in the world,' added Patrick Furlong.[16]

During those times when Parliament was not sitting—known as recess—the Lords Chamber remained unheated except in exceptional circumstances. The furnaces were normally lit and the space aired only about a week or ten days before the House was due to return (which in fact was scheduled for 23 October). This work was usually done by Reynolds the firelighter. He would light the coal in the stoves at two or three in the morning, then tend them during the day. Caution was always required in this work, ordered the Yeoman Usher James Pulman, in a set of instructions to the staff. It was particularly important not to overheat the flues, a subject close to his heart in his role as Black Rod's deputy, he 'having frequently felt the inconvenience of the floor being very much heated near to Black Rod's box where the flues pass'. The flues had never actually set fire to the floor in that part of the Chamber, but it was a touchy subject with Pulman, who remembered a scare some five years previously. One Sunday morning about ten o'clock, Mr Wright the doorkeeper (Mrs Wright's husband) had spotted some smoke issuing forth from the floor. Adam Lee, Weobley's predecessor, had rushed round, removed a skirting board, and found that the wood had ignited in the lobby about halfway between the Grand Staircase and the door leading to the Bar. That accident was caused by another flue—one rising from a stove at the foot of Soane's Grand Staircase (the *Scala Regia*) which led from the King's Entrance up to the Lords Chamber, via Soane's Royal Gallery and the Painted Chamber, on the first floor. Pulman knew of no other occasion when such an incident had occurred.[17] Elizabeth Wright remembered that day too. When fire did break out in the Chamber in the late afternoon of 16 October 1834 she initially thought it was just the matting, 'which has frequently caught,' she said, '—or I should have saved some of my things'.[18]

Later, Mrs Wright appeared 'unusually afflicted' during her questioning by the Privy Council about her role in the events of that fateful day.[19] Asked if the stoves had been known to overheat before by being stoked with too much fuel, her odd response was, 'Yes, and that was my reason that the men should not have a light; I did not see a man have a light in the House of Lords the whole of the day; none of the workmen had any.' She also claimed to have sent word down to the furnaces a further two times—at half past one and four in the afternoon; claims

61

which no other witness corroborated, and which some, in fact, strenu-
ously denied, particularly Jukes, the matting foreman, whom she said
she had sent down at four o'clock. The only afternoon visit which
everyone agreed did happen was that of David Reynolds. Even then,
Mrs Wright said he had gone down at five, but everyone else told the
Privy Council it was between three and four. This must have been so,
as Cross and Furlong left shortly after four, and David himself finished
work soon after that. At no time did the deputy housekeeper see fit to
visit the furnaces herself.[20] Whether her confusion and forgetfulness in
front of the Prime Minister were caused by terror, an inept attempt to
cover up her lack of action, or incipient senility is not clear. Perhaps it
was a little of all three.

The heating of the House was not all it could be. A Lords committee
had considered the subject in 1831, and ordered Sir Robert Smirke (the
Office of Works architect, restoring Westminster Hall at the time of the
fire) to put into effect some plans for improvements he had presented to
them.[21] The mouths of the furnaces in the vault under the Lords
Chamber were in fact only feet away from the base of Black Rod's Box.
Their flues, running under the floor above, were swept just before the
beginning of each session. In October 1834 they were shortly due to be
cleaned of the layers of hardened tarry creosote known as clinker
which build up in every chimney flue as a result of the incomplete
burning of coal or wood. As anyone who lives in a house with an open
fire or solid-fuel stove knows, too much clinker or soot in a chimney
runs the risk of being set alight by sparks or the heat below. The Westminster flues were cleaned regularly every year.[22] The Palace bricklayer
called Hindle was in charge of getting them swept. It was a big operation to clean the copper-lined flues under the Lords Chamber, as it required all the floor matting and bump to be taken up by someone from
the Office of Works. The furnace mouths in the ground-floor vault
were too small even for a boy to get up inside the flues to scrape off the
soot and clinker inside, so instead access was provided through doors
in the floor of the House of Lords Chamber. Once unlocked, these

stove-traps could be lifted to expose the flue running underneath. Soot-holes were then cut into the flue bricks so that the sweep's boy could wriggle inside to do the cleaning. The stove-traps were opened each autumn prior to the sitting of Parliament by the Clerk of Works' men specifically for that purpose. It was almost time for them to be swept again.[23] One of the many ironies of the 1834 fire was that, on 24 July that same year, Parliament had passed an Act forbidding the apprentic-ing of boy sweeps under the age of 10. It had become an offence to compel anyone, apprentice or no, to enter a flue to put out a chimney fire. Sweeps and their boys were forbidden from crying out their trade in the streets, on pain of a fine of forty shillings. The diameter of flues and the sharpness of the angles at bends was regulated, in an attempt to prevent the cleaning boys becoming stuck or injured in those less than 120°, and to allow the brushes through easily. Any complaints by apprentices about overwork or abuse by their masters or mistresses were to be heard by the local magistrates.[24]

Alongside many now-forgotten measures of those years, the Whig ministry had introduced a swathe of reforming liberal legislation which is still remembered today. It tried to dislodge a number of outmoded, cruel, and rickety practices which had built up over previous centuries and were now still clinging stubbornly to the early decades of the nine-teenth century. In 1833, the Slavery Abolition Act abolished slavery in the British colonies. Quakers and members of the Moravian Church were permitted to affirm, instead of swearing an oath which was against their belief. The practice of hanging the bodies of criminals in chains after their death, or passing on their remains for dissection, ceased, as did capital punishment for those returning from transporta-tion before their time.[25] The 1833 Factory Act (known as Althorp's Act after the Chancellor of the Exchequer, who sponsored it) attempted to improve working conditions for children in the cotton (and the follow-ing year, silk) mills of the north and midlands.[26] Like the Chimney Sweeps Act of 1833, the Factory Acts proved difficult to enforce. Last but not least, there was the 1834 Poor Law Amendment Act, which attempted to reform the way in which the poorest in society were supported. It promoted the institution of the parish workhouse as an active deterrent against poverty, and swept away the old Elizabethan system of 'outdoor relief', that is, the providing of payments in money or in kind to the vulnerable in their own homes through the parish (then the principal instrument of local government). Intended as a logical, modern, scientific method of managing the costs associated with supporting the poor in rural areas, in the long term it went disastrously and inhumanely wrong. At the time of the fire—only two months after it had been passed—it was already arousing a good deal of opposition from middle-class anti-Utilitarians, and from the poor themselves.[27] Exactly how much opposition was to become clear during the evening of 16 October 1834.

Shortly after four o'clock that day, Mrs Wright was showing round two curious gentlemen-tourists. Mr John Snell, an ironmonger of Tiverton in Devon, was staying with his friend Mr John Shuter during a visit to London.[28] They began by walking through the House of Lords Chamber, and, on entering it, observed considerable smoke in the air. By way of explanation, the housekeeper offered, 'the workmen are below', but made no further comment. They walked up to the throne in silence and then exited behind it into the Robing Room, where there was also a little smoke in the air, but not much. From there they passed through into the 'Conference Room' (their name for the Painted Chamber) for five or ten minutes.[29] It must have been approaching ten past four when they returned to the Robing Room, where they stayed for around a quarter of an hour, then—on Mrs Wright's suggestion—strolled through to the 'old Lords House' (presumably their name for the Royal Gallery, on the site of the old Chamber). One of the men observed how chilly it was there, with a 'raw air' (perhaps the door from the King's Entrance was open) so they did not stay long—about five minutes—before walking back to the new Chamber around half past four.[30] A contemporary guidebook for tourists informed visitors that this was by no means a splendid room, but it was nevertheless very handsome. The old canopy of state under which the throne was placed remained as it was before the 1801 Union with Ireland, except that its tarnished appearance was rendered more conspicuous by the arms of the United Kingdom being inserted into the old stuff, embroidered with silk, with silver supporters. The throne itself was a large gilt chair, with elegantly carved and gilded arms, ornamented with crimson velvet and silver embroidery.[31] Another guide commented: 'The Speaker has no chair, as in the Commons, but is seated on a large woolsack, covered in red cloth, with no support for the back, nor any table to lean against in front. This is a most preposterous and almost cruel custom.' The peers' benches were covered with red cloth and,

> there is a bar across the house, at the end opposite the throne, at the outside of which sits the King's first gentleman usher, called the black rod, from a wand he carries in his hand. Under him is the yeoman, who waits at the inside of the door, a crier without, and a serjeant-at-mace, who always attends the lord chancellor.[32]

This was all very pleasing for the visitors, and they had a thoroughly diverting time. They stood by the throne for a while, admiring it (Mr Snell even 'got upon the side' of it, to his delight), and inspected the woolsack and other places for about fifteen minutes.[33]

With one part of their visit Mr Snell and Mr Shuter were disappointed, however. That was their experience of the famous Armada Tapestries, which were one of the main visitor attractions of the old Palace. Just like the 1588 naval battle they portrayed, the tapestries enjoyed an iconic status, representing a seminal moment in England's history, when Philip II's Spanish invasion force was repelled by a combination of seamanship, tactics, and favourable weather, thus preserving the Protestant rule of Elizabeth I. They were commissioned by Lord Howard of Effingham, the Lord High Admiral and naval commander responsible for the defeat of the Armada. Howard wanted a lasting memorial to his triumph and ordered the design and engraving of eleven naval charts showing each stage of the confrontation. These

65

were each then woven in Delft by the tapestry master Frans Spiering and his workshop to designs by the celebrated Dutch maritime artist Cornelius Vroom of Haarlem, based on the charts. They portrayed, from a bird's-eye view, each stage of the confrontation in the English Channel from 29 July, when the Spanish fleet was first spotted off Plymouth, to the final battle of Gravelines on 8 August, when it was at last routed and scattered.[34] The tapestries had hung in the Lords Chamber since the middle of the seventeenth century. By 1834 they were no doubt faded and grubby, but still spectacular, resonant with English history, and an important highlight of the Westminster tourist trail.

But their condition was not the reason why Mr Snell was taken aback. It was because his view from the Bar was hazy. The House was filled with smoke. Recalling his visit afterwards, he said, 'I was so much grieved that there was so much smoke in the House that I could not see the Tapestry, and I went over and put up my hand to convince myself that it *was* tapestry; but I do not think I saw above a foot square when I was near to it.' They were about to leave at a quarter to five when their attention was attracted by Black Rod's Box, and they went over to have a look. On standing by the Box, a surprised Snell felt a strange sensation, and the following exchange took place:

MR SNELL: Bless me! How warm it is in here! I can feel the heat through my boots... (*bending down to put his fingers to the floor*) I should almost be afraid this would take fire!

MRS WRIGHT: Oh no, sir, it is a stone floor.

MR SHUTER: (*leaning over into the box*) Mr Snell, step over this way and only see what a suffocating heat there is!

MR SNELL: Yes, there is a very great heat; a suffocating heat!

MRS WRIGHT: Oh, the workmen are below burning the Exchequer tallies, and I have known when the House has been sitting, that people that have stood at this spot have almost fainted.

The alarm of the two tourists was temporarily distracted by this state-ment, and they stood around chatting on other topics for another quar-ter of an hour. But, reflecting on it afterwards, Snell was struck by the strangeness of this encounter. Smoky rooms, caused by chimneys which did not draw very well, were commonplace at the time, but this seemed out of the ordinary. Fumes did not seem to be coming through the floor, however. 'I was never in the House before,' he mused later, 'but I thought it a strange thing that there should be so much heat and smoke.'[35] Towards five they left, and Mrs Wright locked up after them just as dusk was falling. David Reynolds had been sent home about half past four.[36]

But Mrs Wright was wrong to tell John Snell that a stone floor would protect the Chamber from a fire outbreak. There was a good deal of combustible material directly above the flues. Near Black Rod's Box the floor was covered with the oiled bump, to prevent the dust filtering through, as it would through matting. Within the Bar the floor was wooden, the Bar itself made of canvas-covered wood with a covering of scarlet cloth. The sides of Black Rod's Box were again scarlet cloth-covered canvas on wood. The walls of the House below the tapestry were covered in canvas and scarlet cloth, painted to look like open wooden panelling. In the Lobby outside the Bar the floor was wood, and the walls from floor to ceiling were hung with distemper-painted canvas.[37] Surrounding Black Rod's Box was a crimson curtain.[38]

Immediately below, in the vault, it is likely that the copper lining of the flue running under the floor exactly at the corner of Black Rod's Box had by this time begun to deform and collapse. Lining a brick flue with copper was an odd choice in the first place. The heat from the wood fire coursing through it had probably melted the copper nearest to the furnace, while that further away had become distorted. In addi-tion, the repeated holes cut into the flue via the stove-traps in the floor meant that there may have been fissures compromising the brickwork. This would allow heat and sparks to emerge and make contact with any woodwork or matting above the stone floor. It was this that was

67

allowing the smoke to seep from the flues—though it was not apparent to Mr Snell where the smoke was emerging. By this time the heat in the furnaces would have been intense—copper melts at 1084.62°C. What had happened was that Cross and Furlong had unwittingly created a reverberating furnace. Coal (which was what the flues were built for) burns long, low, and slow at 600–800°C with little flame. Dry wood, however, burns very rapidly, especially if the dampers or the doors on a stove are open (increasing the intake of air), and the fire is constantly and ferociously stoked.[39] A fire in a stove becomes fiercer when its vents or doors are kept open, allowing more oxygen in.[40] It is not known what state the workmen left the furnaces in shortly after four. They may have conscientiously watched the final sticks burn down, damped the embers, and closed the stove doors. Or they may have piled the final batches of sticks high in the mouths to burn away and abandoned them without further ado. It probably matters very little, because whatever condition Cross and Furlong left the furnaces in, the real damage was already occurring higher up, inside the inaccessible flues.

Much later, on hearing the news of the fire and discussing it with his friend, Mr Snell remembered the heat in his boots right by Black Rod's Box, and it flashed immediately into his mind that it must have started from that spot.[41] James Pulman, the Yeoman Usher, also had no doubt of the origin of the fire. He knew the furnaces themselves were very nearly directly under the Bar door, just a few steps from the Box.[42] Sir Robert Smirke, called as an expert witness by the Privy Council, was of much the same opinion. If a square, brick flue was built directly underneath combustibles and no more than ten feet above a furnace (which had been filled three-quarters full with very dry wood and then burned with the door open), then he considered the situation described would heat up the bricks of the flue so much that they could indeed set fire to the flammable objects. It would be very much more dangerous than a coal fire burning. He later confirmed the flue as being 14 inches wide and 18 inches deep—its length from Black Rod's Box being about thirty feet and one of the stoves being not less than eight feet directly below it in the ground-floor vault. 'The effects of an extraordinary degree of heat in the flue are seen at a considerable distance beyond the part nearest to the Usher's box,' he reported to the Privy Council inquiry on 21 October. He was asked whether he had been able to examine the flue's construction in detail. 'Not much,' he replied, 'for it was so much destroyed.'[43]

The inquiry also heard from Richard Weobley. Shortly after five o'clock, Joshua Cross left the Star and Garter where he had been drinking with Furlong, encountered his Clerk of Works in the street, told him the burning was over, and returned home for his tea.[44] Weobley simply could not believe what happened subsequently. The burning was not over. 'It appears on consideration,' he said in his evidence, 'that the flues and place altogether can never be of the solidity that I had supposed it to be; I supposed it was the safest place that was possible, and if the flues had been solid and sound there could be no harm.' The flues had been in use for many years but despite this he now realized they might be defective. It never struck him, he admitted, that burning the tallies where he did would have the effect it did. Sitting before the inquiry, he maintained a kind of dazed puzzlement about what had occurred. Surely no quantity of tallies would produce the kind of temperature that would heat the flues round the bricks so as to make the beams beside them burn? Counsel to the inquiry, probably the Lord Chancellor, pressed him as to whether he thought the burning of the tallies might indeed be responsible for the terrible destruction of that night. 'I am very sorry to give my opinion, which may affect myself,' confessed a chastened Weobley. 'But I am very sorry to say that it has very much the appearance of that.'[45]

5

One of the greatest instances of stupidity upon record

NIGHT WAS FALLING. The full moon was near, and a strong breeze was blowing from the south-west, across Abingdon Street and over the west front of the Palace. Occasionally large clouds floated high and bright across the starry sky. The Thames was silent and distant, the tide now at a low ebb.[1] All was quiet. A few minutes past six o'clock, Mrs Mullencamp (the wife of a Lords' doorkeeper and a neighbour of Mrs Wright's who lived inside the Palace 'on the corner') returned from an errand. As she passed through the building, in the gloom she saw a flickering light under the north door of the Lords Chamber where the Commons entered at State Opening. 'Oh good God!' she screamed out, 'The House of Lords is on fire!'[2]

Mrs Wright was with the family in her apartment. She seized her keys and—remembering the previous occasions when the matting had begun smouldering—she ran first to the door by the Bar of the House, and then scuttled to the door opposite the Throne. Opening it up she saw the side of Black Rod's Box on fire 'and the flames going up', ready to catch—or already catching—the nearby red velvet hangings. She ran to the apartment of another doorkeeper, Mr Moyes, to get help. Mrs Moyes swiftly fetched her husband, the receiver of peers' letters, and both he and Mrs Wright hurried back to the Robing Room behind the Chamber.[3]

At much the same time, Weobley was standing at the Abingdon Street corner of the Palace, seeing off some men who had been working on the enlargement of the House of Lords' Library. The official duties of the Clerk of Works being finished at six in the evening, Weobley turned to go home. Something caught his eye on the rooftops against the night sky and, startled, he saw the first exterior signs that all was not well. A chimney was 'very much a-fire', though he could not see which one. It might have been one of the flues, or the chimney of Mr Howard's coffee-house: it was impossible to tell—they were all in a stack. He ran up the street to the other end of Old Palace Yard, to the stairs leading to Mr Bellamy's refreshment rooms at the very top, dashed up them into the House of Commons, and then turned to enter the Lords Chamber. Despite the fact that the Lobby to the Chamber was in darkness, the glow beyond it was unmistakable: 'I saw then under the door that it was all on fire', he later gave in evidence. Turning to see Mr Moyes and Mrs Wright coming round the corner, he wailed, 'Oh Lord! It is all on fire! Take care, for it is all on fire! It is much more than a chimney!' He never forgot the faces of the other two at that moment. Running down the stairs as fast as he could to get to Cotton Garden, and now in a great panic, Weobley yelled, 'Fire! Fire! Fire!', rushed through the passage towards Mr Ley's house (past the furnace vaults), round the side and into the garden, and then climbed Guy Fawkes' staircase, on the outside wall of the Painted Chamber, thinking that he would be able to get buckets there to extinguish the fire. Once at the top of the stairs it was apparent that the fire had already taken hold and was spreading so rapidly around the chimney that he could not get on to the roof at all.[4] Mrs Wright ran back up to her apartment to warn her servant, afraid she would be killed.[5] By the time Weobley ran back down the stairs, a crowd of his staff had gathered, and he got

someone to run round to the Speaker's House to warn the Commons.[6]
Still no one outside the Palace knew of the calamity unfolding within.

Only a speedy and effective response from the authorities, aided by
well-equipped and coordinated firefighting, could now prevent a national catastrophe from occurring. Since the first half of the seventeenth century there had been fire pumps in London capable of
spurting water, pumped by hand, but without the necessary balancing
mechanism to ensure a steady stream. Following the Great Fire of
London in 1666, a variety of fire insurance companies sprang up, each
with its own trained firefighters, who wore characteristic, brightly coloured livery. Householders and businesses would indicate the validity
of their insurance policy by attaching a metal plaque to the wall of
their premises, displaying the insignia of the company which was easily
validated by the firemen on arrival at the scene. Ingenious firefighting
equipment now included such inventions as the 'cyclo-elliptical pump'
and the 'new sucking worm engine'. The engine consisted of a tank
manually filled with water from buckets. The water was then pumped
out by hand-cranked levers through fixed nozzles on the sides of the
engine. This gave way in time to an attachment for drawing water into
the tank from a nearby source, with nozzles to shoot the water out
through flexible leather hoses, allowing a far greater range for the
stream of water and meaning that both engine and pumpers could be
some distance from the fire. By the middle of the eighteenth century,
London was famous for its firefighting inventions. Of particular note
were those of Richard Newsham, 'the father of fire engines', whose
hand pump could shoot 110 gallons in 60 seconds, with enough force to
break a window. Nevertheless, firefighting was incredibly hard manual
work. While the firemen directed the hoses and fought the blaze, the
mechanism was managed by an engineer, while teams of labourers or
passers-by had to push the horizontal bar on either side of the engine
up and down in order first to draw the water into the engine and then
to force it out down the hose. It was also possible to climb on top of the
tank to pump the mechanism from above, but both techniques

depended on a nearby source of water to keep the tank topped up through an inlet hose. Five minutes was the maximum time tolerable for a stint, so pumpers were constantly being rotated in and out, with beer on hand to keep them from getting dehydrated, and also to encourage onlookers to help out. Crowds would rhythmically chant, 'Beer-oh! beer-oh!' to set a pace for the pumping and to spur on the teams.[7]

There was also a two-hundred-year-old Act in force which imposed some national standard for fire prevention. The Parish Pump Act of 1708, 'for the better preventing of the mischiefs that may happen by fire', required that all new houses should be built with front and rear walls of brick or stone; and between each house a party (that is, separating) wall was to be installed, of two bricks' thickness in the cellar and 14 inches thick above ground. There were to be no timber eaves. Importantly, each parish was to have and maintain a large horse-drawn engine, a smaller hand-drawn engine, and a leather pipe for filling the engine without buckets. Stopcocks were to be fixed to each water pipe in the parish and their positions marked. The waterpipes were made of wood and by the nineteenth century ran under the streets. Officials known as turncocks were to be paid one shilling for opening these 'street-plugs' to provide water for firefighting. And as soon as fire broke out, all parish constables and beadles were to attend with their engines, with incentive rewards for the second and third engines arriving on the scene with a working pipe.[8] Quite how effective these parish officials were, and to what extent they had become figures of fun by 1834, is evident from this vignette:

> Such are a few traits of the importance and gravity of a parish beadle— a gravity which has never been disturbed in any case that has come under our observation, except when the services of that particularly useful machine, a parish fire-engine, are required: then indeed all is bustle. Two little boys run to the beadle as fast as their legs will carry them, and report from their own personal observation that some neighbouring chimney is on fire; the engine is hastily got out, and a plentiful

supply of boys being obtained, and harnessed to it with ropes, away they rattle over the pavement, the beadle, running—we do not exaggerate—running at the side, until they arrive at some house, smelling strongly of soot, at the door of which the beadle knocks with considerable gravity for half-an-hour. No attention being paid to these manual applications, and the turn-cock having turned on the water, the engine turns off amidst the shouts of the boys... We never saw a parish engine at a regular fire but once. It came up in gallant style—three miles and a half an hour, at least; there was a capital supply of water, and it was first on the spot. Bang went the pumps—the people cheered—the beadle perspired profusely; but it was unfortunately discovered, just as they were going to put the fire out, that nobody understood the process by which the engine was filled with water; and that eighteen boys, and a man, had exhausted themselves in pumping for twenty minutes, without producing the slightest effect![9]

This satirical piece was written by a young Parliamentary reporter shortly after the fire. His name: Charles Dickens.

At 6.30 p.m., the immediate neighbourhood of the Palace was thrown into 'an extraordinary consternation' and 'the utmost confusion and alarm' by what was clearly more than just a simple chimney fire.[10] A 'gigantic volume of flame' suddenly exploded from the northern end of the frontage of the House of Lords—directly opposite Henry VII's Chapel and close to the corner with Westminster Hall. The flames simultaneously burst out from the roof over Howard's coffee-house, and filled the ground-floor passage leading to Cotton Garden, and the vault under the House of Lords: just where the furnaces lay, in fact. It was clearly a terrifying spectacle at this stage: burning, 'with a fury almost unparalleled'.[11] The Mullencamps fled, Mr Mullencamp being 'obliged to make his escape without waiting to put on his coat'.[12] Mrs Wright ran heedlessly into the street without her bonnet or shawl, followed by her maid, with the melting lead from the roof already pouring onto her shoulders.[13]

One of the watchmen by Howard's coffee-house—who had previously been oblivious to the burning chimney above him—declared that the fire must be the result of an 'explosion of gas'.[14] Though the initial report by the London Fire Engine Establishment clerk the day after described the supposed cause of the fire 'not known', it did note the possibility of gas being present.[15] To onlookers it was certainly unclear whether there had been an explosion from the gas lighting in the Palace or if in fact the fire was due to the works going on to repair the Bishops' Entrance and apartments, and the Lords' Library.[16] The fire bursting out of the roof led to more than one commentator suggesting there had been a kitchen accident in Howard's coffee-house, which occupied the corner angle between the Lords and Commons.[17] Another was struck by the nature of this first flash of powerful flame. It was (from Lambeth in the east) a peculiar colour, 'totally unlike the usual hue of fire, the tails being of a light blue similar to that of inflamed spirits of wine', and purplish 'like that from turpentine'. It seemed suspicious to him, but soon it settled into the usual colour of a raging fire.[18] Further away, far to the north of Regent Street, there was a different experience: 'so suddenly did that brisk red flame burst from the dark and jagged line of the horizon made by the house-tops, that it seemed rather the instantaneous eruption of a volcano, than the usual progress of an urban conflagration'.[19]

Flashover is the phenomenon which occurs when a fire reaches a particular temperature in an enclosed space. The hot smoke and gases being released by the fire rise to the ceiling, where they can reach up to 1000°C, and then start to heat the space as they fall. When the rest of the room arrives at a certain temperature, everything ignites, causing a rolling ball of explosive flame to consume all in its path. It is terrifying, destructive, and immense. The descriptions of the explosion bursting out of Wyatt's west front and through the roof at around 6.30 p.m. were probably of a flashover (or several flashovers) from the fire within the Lords Chamber, the passage in front of it, and maybe the network of chimney flues themselves. Oxygen entering the Chamber through the opened windows would have accelerated events. The blue flame was characteristic of burning copper.[20]

By the time the alarm was first raised by people in Palace Yard about 6.35 p.m., a considerable portion of the House of Lords was alight.[21] How very curious, people thought, that the fire should have got to such a stage, and that it should have been left to strangers to raise the alarm.[22] The quickest engines on site were not from the London Fire Engine Establishment. The first response in fact came from the parish authorities of St John's, Smith Square, very close by, whose engine was instantly followed by another from the County Fire Insurance Company, one of the insurance firms which had refused to be absorbed within the LFEE. These were driven to immediately in front of the House of Lords and prepared for action.[23]

Burning buildings always attracted spectators, happy to watch and comment on proceedings. Some bystanders felt at first there was a great want of water, and the engines had very little effect at this stage.[24] Infuriated by this slur, the Chelsea Waterworks Company later protested strongly at the reports in the papers that there was an indifferent supply of water available for such a huge fire, though amply sufficient for any normal occasion. Eleven sworn depositions of firemen and turncocks backing its claims were made before the Queen's Square magistrate, Mr Gregorie, and provide a unique insight into what happened in the chaotic minutes after the alarm was raised.[25] Nine street-plugs (or today, fire hydrants) were in use that night, of which it is possible to identify the location of at least five. There was one by the Star and Garter public house, to the south; one at Poets' Corner between the Abbey Chapter House and the Henry VII Chapel serving the western front of the House of Commons and the Law Courts; one opposite the House of Lords; one near Mr Ley's House in a yard providing water on the eastern flank of the Palace; and one in New Palace Yard outside the great door of Westminster Hall, to the north.[26]

William Murless, of Horseferry Stairs, was the keeper of the St John's parish engine. He was alerted at 6.30 p.m. by Henry Butler, one of the messengers of the Lords (one wonders what form that conversation took). We cannot be sure, but the timing suggests that Murless may already have been on his way before the flashover announced the fire's presence to the outside world. Murless kept the Waterworks' tools with his engine, and on arriving at the scene, made for the plug at Poets' Corner. While Murless was screwing the hose onto the engine, James Palmer, a labourer with the Waterworks, turned up, took Murless's crowbar, and used it to draw the plug there at 6.40 p.m., watched by the deputy turncock, James Knight. The pressure of the released water created a fountain as high as Palmer's shoulder (at least four feet) when it first shot out. After the initial release the water dropped to a steady three-foot geyser. Once the dam was made from which the gushing water could be sucked up by the engine, the pumping started; ten minutes later and the firemen were standing half leg-high in water, and the engine had to be shifted onto higher ground as there was so much water flowing about. Murless oversaw the working of the St John's engine until seven the following morning, using the Poets' Corner plug throughout.[27]

Edward Burch, engine keeper to the Globe Insurance Company, was called to the fire at 6.40 p.m. and also went to the Poets' Corner plug. He found 'a small engine' there (presumably St John's) and plenty of water. Once he had made a hole in the dam for his hose to suck, he set the engine to work until about 9 p.m., and then moved round to use the Star and Garter plug, at which point there were eight engines all feeding off Poets' Corner. A fireman for thirty-one years, 'he never saw so much water at a Fire in his life'.[28] James Rooke, foreman to the County Fire Office, residing at the engine house in Marylebone Street, arrived at Poets' Corner at much the same time. The County was the first horse-drawn engine on the scene, and Rooke was there till 6.40 p.m. on Sunday evening. He too found no want of water and in fact had to pull the engine out of the water as the men were up to their knees and could not stand in it.[29] Also early on the scene at Poets' Corner was the British Fire Office engine (again, a private concern). Charles White, its engineer, had received the alarm at the same time. He ran down immediately, followed by his engine and then, about five minutes later, the firemen. One of his firemen, Edward Saunders, made a barrier with boards and stones to keep the men dry who were working the engine. Water was running through the gratings of the plug into the sewer, and no dung being available to block it up, he got a piece of old carpet and jammed in some straw to block it, and started it working on the Law Courts opposite Poets' Corner. The British Fire Office engine pumped there till one in the morning 'with an abundant supply of water' and was then shifted to the Members' Entrance of the Commons. Saunders, a fireman for thirty-five years, considered the water supply to be 'excellent'.[30]

After having opened the Poets' Corner plug, Palmer the labourer went round to where the St Margaret's parish engine was setting up. Its keeper, Richard Owen, lived at 13 Canon Row, and—like the St John's engine keeper—was called at around 6.30 p.m. before the flashover. He had pulled his engine into Cotton Garden. Palmer opened the iron flap of one of the private tanks but had no tools to turn the cock since it was a private plug of a non-standard design. Within about three minutes, however, Hubble, the principal turncock, had arrived and turned on the water, being obliged to break the cover of the tank to do so. Owen worked for about an hour after 6.40 p.m., until the fire there was burnt out. The yard was knee-high with water when he left and it had served two parish engines during that initial hour. With the assistance of Robert Evans, a fireman from the Norwich Union fire office, the St Margaret's engine was taken through a window in the Speaker's House and placed in the garden. Owen spent the rest of the night working the engine in the Speaker's Garden alongside the other parish engines. He finally took the engine back to its home at five the following morning.[31]

Joseph Hubble, principal turncock to the Chelsea Waterworks, arrived at the fire like all the others at about twenty to seven. He remained busy for the rest of the night. Initially he opened seven plugs and firecocks, and was at the fire the whole night regulating the cocks, with his deputies Thomas Conboy and James Knight. Some plugs had more water in than others, owing to the number of engines at work: some had five or six to a plug, while others had only one or two. Altogether there were nine plugs and firecocks open, and twelve engines at work before 7 p.m.[32] Knight drew the Star and Garter plug serving Cotton Garden.[33] There was a hiccough when one of the two engines working from it stopped early on, but it was got going again after ten minutes, and the engine masters could have had water from a dam about twelve yards further off but would not leave their stations. The gratings at the Star and Garter plug were blocked with some dung from Little Abingdon Street, which in total provided plenty of water for six engines.[34] The place in fact was swimming with water in parts, making it hazardous for people and possessions alike.

The Office of Woods and Forests owned at least five fire engines to look after the safety of royal residences and other important state buildings in London. There was one each at Windsor Castle, Kensington Palace, Hampton Court Palace, Somerset House, and at the British Museum. A good fire engine could cost up to eighty guineas. Within the previous two months, the engines at Hampton Court and elsewhere had been tested, and it was the rule that an engineer checked and repaired them, and cleaned the hoses, once or twice a year. Just the day before, a Mr Simmons was testing the fire engines at Horse Guards, and those particular ones also seem to have been in Old Palace Yard before seven o'clock, helping with the initial efforts.[35] No doubt passers-by were already lending a hand: one of them, William Chandler, a fireman formerly of the Guardian and Palladium fire office, but not attached to any at the time of the fire, claimed to be the first fireman on site, as he lived on the spot and was walking by when the fire burst forth. He helped both the government and parish fire engines to get going.[36] The parish engine from St Martin-in-the-Fields was now also on its way. Four men, Wing, Hill, Wood, and Seaton, brought it to the scene under the direction of its keeper, Mr Dobson, and continued with him there all the night.[37] Despite all these early efforts, in which the firemen exerted themselves to the utmost, the flames had already got so strong a hold that the engines had very little effect.[38]

With flames shooting from the roof of the Palace, the blaze was immediately evident across the Thames in Lambeth. It was quickly recognized as 'one of the most terrific conflagrations that has been witnessed for many years past...the ill news spread rapidly through the town, and the flames increasing and mounting higher and higher with fearful rapidity, attracted the attention not only of immediate passers-by but thousands of persons from across the metropolis who in a few minutes were seen hurrying to Westminster'. *The Times* correspondent who wrote this could not recall ever having seen the thoroughfares so thronged before. Within less than half an hour, it was becoming impossible to approach the foot of Westminster Bridge on the Surrey side, or the end of Parliament Street on the other, due to the hordes of people trying to get across or onto it. The crush could only be avoided by means of a boat or with the assistance of a guide who, well acquainted with the localities, could navigate round the back of Westminster Abbey and reach Abingdon Street at the bottom of Old Palace Yard that way. That space, however, was also quickly filling up.[39] Crowds were gathering close to the buildings in Old Palace Yard, eager to get the best view.[40]

Of the many medieval institutions swept away at the end of the 1820s and early 1830s, the Houses of Parliament were not the only London icon replaced by another which has since become recognizable the world over. In 1829, the same had happened to the Parish Constable. The new Metropolitan Police force—a civil one, rather than military—had been set up under the joint command of two commissioners, known as 'magistrates'. One, Charles Rowan, was a retired Irish colonel in the 52nd Regiment; the other was Richard Mayne, a barrister, also Irish.[41] They divided London into seventeen letter-coded sub-divisions (Westminster was B division), controlled centrally for the first time in its history. The new headquarters was at Scotland Yard, 4 Whitehall Place, a few minutes' walk north of the Palace (next door, in fact, to the Office of Woods and Forests). By 1830 the force numbered about 3,300 men.[42] The subdued blue uniforms of the new

Metropolitan Police, with their white buttons, reinforced stove-pipe hats, rattles, and truncheons, earned them the nickname of 'raw lobsters' (as opposed to the military, who wore red).[43] A few minutes after a quarter to seven, fifty Grenadier Guards marched to the spot and with the assistance of a strong body of police, opened up a square space before both Houses, in order that the firemen were not obstructed in their efforts. Another fifty Coldstream Guards, unarmed and casually dressed, were ordered to help the firemen out.[44] Only a few years before, street brawls between the raw and boiled lobsters had been common, as the new civilian force sought to establish its position alongside the existing forces of law and order.[45] On this night, though, there was no such trouble, with police and military cooperating to control the crowd, man the pumps, and clear the way for rescuers over the course of many hours. Access through the cordon was provided to anyone who could be of help to the firemen.[46] A crude engraving of the scene captures what Old Palace Yard looked like at this stage. Ranks of soldiers are keeping the crowds back from the square, lined with gas lamps. In the wide empty space in front, five fire engines can be seen, but only one is in use. A pair of horses, presumably used to pull one of the engines to the site, shy in fear. Smoke and flame billow from the west front of the House of Lords.[47]

Captain William Hook of the Royal Navy was in the vicinity when the fire broke out. He arrived in Old Palace Yard at much the same time as a detachment of armed guardsmen—probably the Grenadiers mentioned above. Thinking his experience might be of some help, he made himself known to their Commander, a man called Lindsay, and was allowed to pass through the line of police and soldiers, as soon as the latter had taken up their positions. The lack of organization in the firefighting, which was one of the main reasons for the creation of the LFEE, was already plain: each engine crew was accountable only to itself. Hook was staggered to discover (as might be expected from a navy officer) that there was no one coordinating the rescue efforts at the time, not even the police. He repeatedly suggested to the firemen from the local parishes and surrounding government buildings that they should place the engines between the flames and the offices around Westminster Hall, to prevent the buildings 'yet untouched by the fire' from being damaged. But he found that there was a 'total disregard to the general good' and that they merely offered a 'deaf ear to every suggestion made by myself and other gentlemen and continued to pour the stream uselessly into the part of the building already past recovery'. He and others then desperately appealed to the police, but their answer was that they were there only to keep order and had no authority over the fire-engine men. Deeply frustrated, Hook then set about devising a separate strategy of his own, of which more later.[48]

Meanwhile, sitting quietly at home after dinner with his wife, was John Cam Hobhouse, Lord Broughton. 'This, to me, has been a memorable day', he wrote in his diary; something of an understatement coming from the first Commissioner of the Office of Woods and Forests. Receiving the news that the House of Lords was on fire, he swiftly arrived on site. The sight which greeted his eyes when his carriage pulled up was by then highly alarming, and chimes with Hook's observations: 'The whole building in front of Old Palace-yard was in flames, and the fire was gaining ground towards Bennett's Tower every instant; only a few soldiers and policemen were present, and three or four engines. A short time after more fire-engines came, but I thought the firemen were lamentably deficient in the knowledge of the best way to extinguish the flames.'[49]

In their house in New Palace Yard, John Rickman, the Clerk Assistant of the House of Commons, and his wife were having a nap after dinner (they had eaten separately that day, Rickman having dined at his club).[50] Rickman was second only to the Clerk of the House in the Commons' administration, and his house was a large and handsome brick block straddling the space between the north-east corner of Westminster Hall and the Exchequer Offices on the east side of New Palace Yard. While they were dozing, a maid came in to tell their daughter Frances that 'a small fire' seemed to have broken out at the Lords' end of the Palace, and they went up onto the roof to view it. It was just gone twenty past six, so what they could see at first from the slates was just the single chimney burning. But as the two women watched, they saw the first blast of the explosion ten minutes later—'the flame was seen to ascend ten yards above the roof of the House of Lords'—and then the fire began blowing fiercely towards the house of their friend Mr Wilde, Keeper of the Exchequer, and his family, overlooking the river.[51]

Also sitting quietly at home at this hour was Joshua Cross. He had finished his evening meal and was nursing one of his children, when he heard people running past his house. Learning there was a fire, he made his way to the main street and only then discovered that the Palace was 'all in a blaze there'. He later claimed it never crossed his mind that the burning of the tallies might be to blame for the calamity in front of his eyes, but then changed his story and said it might have struck him as *just* possible. Patrick Furlong was having his tea when Cross turned up telling him there were fires round about Palace Yard, and they both ran there together. Furlong was in such a state that he later claimed not to remember what he might have said to Cross at the time about the cause of the fire: 'a man that is in a fright cannot say what he said.' But Cross remembered all right. 'I said I hope to goodness it was not me burning the tallies...he hoped the same...he hoped not...we both ran as fast as we could, as soon as ever I heard where it was, and used the greatest speed to render all the assistance I possibly could, which I did to the utmost of my power two days and two nights.'[52]

Not once during the day did the complacent Mrs Wright go to look at the burning furnaces herself—'alarmed, as she states herself to have been,' said the Privy Council inquiry report, 'by such manifest indications of danger'—nor raise her concerns with Mr Weobley, the Board of Works, or the Home Department. Weobley, in turn, was also later condemned by the inquiry into the disaster for 'not more effectually' superintending the furnace operations, particularly after three o'clock, when the 'symptoms of danger...thickened and increased until the event took place'.[53] Less than half an hour after the fire was spotted in the street, the whole interior of the western flank of the House of Lords could be seen, through the windows, to have become 'one entire mass of fire' from floor to roof.[54] The blaze, overtaking Wyatt's Cotton Mill, 'cast its lurid glare far over the horizon—lighting up the broken clouds above, which were driven on before a strong westerly gale—spreading over the silent Thames a vast sheet of crimson, that seemed to smother the more feeble rays of the rising moon—bringing out the stately and majestic towers of the Abbey in a strong relief against the deep blue western sky, with a bright white tint, strange and sepulchral, in contact with all around—and playing with seemingly wayward and fantastic scintillations on the inimitable fretwork of the Seventh Harry's Chapel'.[55]

Lord Melbourne, the Prime Minister, was at dinner with Lord Althorp and Lord Auckland when they heard the news from Downing Street. Melbourne had become Prime Minister in July 1834, following the exhausted Lord Grey's resignation as the head of the Whig ministry over Irish tithes. Like a number of other onlookers, the experience of the 1834 disaster permanently affected Melbourne's already depressive mental state. In later years when lonely, stressed, or weighed down with a sense of responsibility for the country, he suffered flashbacks in which 'the image of the fire frequently presented itself'.[56] Initially though, the new Prime Minister's response was to the point. It was, he said to his predecessor Grey,

> One of the greatest instances of stupidity upon record: I have no doubt it had been burning the whole day. No private house would have been destroyed in such circumstances.[57]

6

The brilliancy of noonday

SUPERINTENDENT JAMES BRAIDWOOD received the call to the fire at the London Fire Engine Establishment Office HQ at 68 Watling Street in the City, east of St Paul's, at about seven o'clock.[1] Braidwood, a Scot, was the son of an Edinburgh cabinet maker and builder. Originally a surveyor, he was appointed first master of the fire engines of the newly formed Edinburgh Fire Establishment in 1824, at the age of 24. From surveyor to fireman seems a strange career path today, but at the time firefighting was thought to be particularly the preserve of those who understood building structures and materials, and who had no fear of heights. Sailors were also popular recruits on account of the latter requirement, and also their fitness and bravery. Braidwood had found his vocation. He developed an efficient force in Edinburgh, imposing discipline and weekly fitness drills at four in the morning on his men, and logically dividing the city into districts the better to respond to emergencies. These innovations were extremely effective, and his pioneering firefighting methods, on which he published widely, included the principle of directing water at the base of

85

the flames, and instructing that firemen—in a particularly British development—should enter burning buildings under the command of their officers, to fight the source of the fire. His invention of the chain ladder to rescue people trapped in burning buildings won him the Society of Arts' silver medal in 1830.[2]

Two years later he was headhunted to lead a brand-new firefighting force in London. By 1832 the increasing number of fires in the city, a drop in discipline, a decline in the number of engines from fifty to thirty-eight, and a deterioration in the condition of the remainder, meant that the insurance companies there were in crisis. On 1 January 1833, ten of them pooled their resources and created the LFEE—the London Fire Engine Establishment—a consolidated, private, metropolitan fire association for London.[3] Braidwood was appointed its Superintendent, charged with unifying the previously competing units—the Sun, Protector, Imperial, Westminster, Atlas, Alliance, London, Royal Exchange, Union, and Globe insurance company brigades, which were joined later by the Phoenix and the Guardian. Among those who kept aloof were the British, County, and West of England companies.[4] The LFEE comprised fourteen engines located at thirteen day and night stations across the capital, manned by five foremen to supervise operations, nine engineers to maintain the engines, and sixty-six firemen to control the hoses and direct them against the flames. The engines were driven to the sites of fire by horses and drivers hired from local car-men. On call throughout the night via a speaking tube in his bedroom at the Watling Street HQ (which was constantly lit by gas to aid a rapid response), Braidwood's new techniques were now to be put to one of their greatest tests. Someone had seen the light in the air to the west. He immediately rushed to the Palace with one of the Watling Street fire engines, while the others converged on Westminster from across London.[5]

The appearance of the fire from the south-western end of the Palace, at the corner of Abingdon Street, was now 'exceedingly striking' and, for a time, the non-LFEE firemen worked on trying to save the tower that rose above the Peers' Entrance at the end of the west front, which housed the Lord Chancellor's office on the principal floor. All the rest of the frontage of the building was enveloped in flames, now extending themselves right along the Abingdon Street façade overlooking Old Palace Yard. On the upper storeys here were the Lords offices, while on the floor below were the stone steps leading to the House of Commons.[6] The fire was so centrally placed in the Palace that it had ready access not only to Lords buildings but also—lethally—to the nearby passages and lobbies of the Commons. In the hours that followed, as one room or corridor caught fire, the toxic smoke produced ran ahead of the flames, slithering its way under doors, insinuating itself through floorboards, and snaking steadily up staircases, heating each area as it went and preparing them to ignite. Throughout the night, the blaze was to spread in three directions—along the west front of the Lords; eastwards beyond the Lords Chamber to the river front, where the Painted Chamber, Speaker's House, and both Libraries lay; and northwards towards the Great Hall and St Stephen's. Observers noted how 'the rapidity with which the flame proceeded, proved that not only had the fire got decided hold of the premises, but that it threatened the whole of the buildings, which were so connected one with the other, as to endanger all if the fire once got the mastery, and so to afford no chance of intercepting its progress'.[7]

At ninety degrees to the House of Lords façade, and jutting out towards the apse of Westminster Abbey, were parts of the Law Courts building belonging to the House of Commons. This was taller than the rest, the upper part being a portion of Bellamy's refreshment rooms, and the lower being used as the MPs' cloakroom. For some time, this wing, like the octagonal tower above the portico over the Peers' Entrance, was hardly touched by the flames, and seemed 'to bound the ravages of the fire and to offer successful resistance to its further progress, while all between this was in one uninterrupted blaze', attracting universal attention. 'The flames in fact extended beyond these two points, but seemed to exhaust themselves in the destruction of them. They took fire nearly at the same moment, and burning furiously for nearly half an hour, the whole structure, from the entrance of the House of Commons to the entrance of the House of Lords, presented one bright sheet of flame.'[8]

Terrifying sparks were now falling 'thick as flakes of snow' on Pedlar's Acre, north-east across the river in Lambeth (today the site of the former County Hall). By half past seven it was possible to read the small print of a newspaper there by the light of the flames nearly a third of a mile away. Onlookers in Old Palace Yard stared at the scene in 'mingled awe and admiration'.[9] Despite the strong breeze blowing, the heat from the flames was enough to keep them warm at that hour and throughout the night. Fire appeared to be bursting out of almost every window of the western façade of the Lords, as well as breaking through its roof, shooting up to an immense height. Some said later that the flames were so bright they obscured the light of the moon, and contrasted with the general blackness of the sky.[10] Most, however, agreed that the moon shone 'with great brilliancy' over a sky already tinted with crimson. The red sky could be seen 'even in the close streets of the city', signalling far and wide that a great fire was under way.[11] This made more of the curious converge on Westminster, and the first rumours that Westminster Hall and even the Abbey itself was burning began to circulate, attracting yet more onlookers.

Westminster Bridge was already so crowded that it had become almost impassable. Its walls and balustrades on both sides were crushed with spectators. The owners of various wagons, carts, and small carriages were enterprisingly hiring out their services to carry richer spectators across for small gratuities, enabling them to get a better view of the spectacle.[12] Gentlemen's coaches and hired hackney carriages were somehow being driven to and fro and, given the size of the crowd, 'another glaring instance of the reckless conduct of those ruffians, cabdrivers, presented itself', *The Times* correspondent sniffed.[13] A passenger in a cab asked his driver about the cause of the fire and was told some were saying that builders did it to create work for themselves, set on by Joseph Hume, 'cause you see sir, the members wouldn't build a new house, though Mr Hume has ax'd'em ever so many times to do it, and told'em how wery uncomfortable he was in the old'un'.[14] Other bridges up and down river—Vauxhall, Waterloo, and Blackfriars—began to fill up. From there, the source of the fire could clearly be identified. The Lambeth shore opposite became thronged with people watching the fire grow steadily behind the buildings on the Palace's eastern flank. Below those buildings—by the Speaker's Garden and at the east end of St Stephen's—grew a crowd of thousands of men and boys, and some women, ankle-deep in mud, as it was low water.[15] And the river was just as busy, with innumerable boats of all shapes and sizes pressed into service, 'which were enabled, by the vivid light cast on the water from the burning pile, to navigate with as much safety as by the light of a summer's sun at noon day'.[16]

Every street and lane leading to the Palace was 'thronged by an almost countless multitude of all descriptions of persons'.[17] Old Palace Yard, Parliament Street, and Abingdon Street were packed out, with the army and police continuing to keep the hordes well back from the blazing building.[18] The Abbey opposite was 'lit up to its highest pinnacles by the awful light'.[19] The flames rising from the buildings were quickly visible many miles away on high ground above the city, in the neighbouring villages of Hampstead, Highgate, and Blackheath.[20] Some simply took advantage of the proximity of the fire to their own homes to climb to the upper floors and view it from there, for many hours.[21] Melbourne the Prime Minister, Duncannon the Home Secretary, and Althorp the Chancellor of the Exchequer had by now arrived on the scene, and were anxiously watching the progress of the flames.[22] David Gregorie, the magistrate of Queen Square (the main court in Westminster), was there, having dispatched all of his officers to help wherever they could.[23] Lord Hill, in command of the army, soon arrived, as did Rowan and Mayne, the twin commissioners of the Metropolitan Police.[24] At some point too, John Phipps encountered his Clerk of Works, Richard Weobley. No record of their conversation at that moment survives.[25]

The Prime Minister spent most of the evening with Lord Broughton, the first Commissioner of Woods and Forests, and together they ordered the evacuation of government papers from the Law Courts and Exchequer. Horse-drawn vehicles of all kinds began to be requisitioned to take the public records to safety in private houses nearby and to the State Paper Office in Downing Street. The Grenadiers ('in their undress') not only helped to control the crowds, but also to work the fire engines. It was they who began removing the public records—about seven or eight 'large caravan loads', escorting the wagons, when filled, to places of safety.[26]

The Armada Tapestries on the walls of the Lords Chamber must by this stage have completely perished. If they had not initially been burnt by the fire spreading from around Black Rod's Box, they would certainly have spontaneously ignited in the flashover. To get a sense of what was lost, it is important to understand the place they had held in the popular imagination since they were created. The tapestry workshop of the Delft master Frans Spiering in the sixteenth century was the equivalent of the most luxurious European design houses today. It produced some of the most expensive Continental pieces of conspicuous consumption of the late sixteenth century. The ten tapestries, when finished in the 1590s, each measured, on average, a huge 24 ¼ feet (7.39m) wide by 14 ½ feet (4.42m) high. The coloured yarn used was a very high quality wool and silk mix. Woven into the designs were gold and silver threads (thin strips of beaten gold and silver wound around skeins of silk), highlighting the crests of waves and other features to give a three-dimensional effect. Each tapestry was surrounded by a wide decorative border containing swags of fruit and flowers, animals, and cherubs, interspersed with medallions containing portrait heads of the commanders who won the victory, and featuring Howard of Effingham's arms in the top corners and his portrait appearing in the centre above the royal arms. After their completion in 1595, Howard hung the tapestries in his manor at Chelsea before moving them to his new residence, Arundel House. In 1616, he sold them to James I for the gigantic sum of £1,628, just over £100 more than he had paid for them.[27]

As part of the royal collection, the tapestries were sometimes brought out of storage to provide decorations for weddings and Christmas festivities, and were first used as hangings in the House of Lords in 1644. They even survived the sale of Charles I's goods following his execution in 1649, and then became a permanent fixture at Westminster. For nearly two hundred years they decorated the walls of the House of Lords, in the process becoming a silent commentary on the political debates taking place in front of them about liberty and the defence of the realm against hostile European neighbours. By the end of the eighteenth century they had become completely iconic. They appeared in John Singleton Copley's famous painting of *The Death of Lord Chatham* (now at Tate Britain), depicting the fatal collapse of Pitt the Elder in the Chamber in 1778 just as he had insisted on the need to continue war in the American colonies, despite the risk at home from invasion by France and Spain. And in a satirical cartoon of 1798 entitled *Consequences of a Successful French Invasion*, James Gillray drew the tapestries being slashed and burnt on the instruction of a French admiral at a time when fear of invasion from France was high.[28]

In 1801, the Armada Tapestries were cleaned and carefully moved when the Lords Chamber relocated to the Lesser Hall. At that time they were enclosed in large frames of brown-stained wool, and hung flat against the walls.[29] The new Chamber would have accommodated eight of the tapestries, with the other two perhaps in the Robing Room behind the throne. They seem to have been in a degraded condition by then, possibly with damaged tops due to improper hanging over the decades, which necessitated the frames. They were known to be faded and dingy by the time of the Reform bill debates, but still atmospheric and highly evocative.[30] In 1739, John Pine the engraver had captured the designs in a plate-book called *The Tapestry Hangings in the House of Lords representing the several Engagements between the English and the Spanish Fleets*. Today, those plates are the main source for what the tapestries looked like. Pine was acutely conscious when he engraved them that, as

91

he put it: 'Time, or Accident, or Moths, may deface these valuable Shadows', and he was right to suggest that by producing the books, knowledge of the tapestries, 'by being multiplied and dispersed in various Hands, may meet with that Security from the Closets of the Curious, which the Originals must scarce always hope for, even from the Sanctity of the Place they are kept in'.[31] Between seven and eight o'clock in the evening of 16 October 1834—the sanctity of the place they were kept in having been defiled—all that was perhaps left of these fabulous works of art were myriad tiny globules of melted gold and silver, burnt away from the textile, and glittering in the hot ashes on the stone floor.

Close by, in the Commons offices, a number of clerks had imagined that storing valuables in their offices in the Palace would afford them better protection than at home. The building, so ancient and familiar, must have had an aura of impregnability. Their faith proved unfounded, and the speed of the fire rapidly overtook their offices in the next hour or two. John Dorington, Clerk of the Fees, and John Bull, Clerk of the Journals, were executors of the will of their late colleague William Gunnell, the Clerk of Ingrossments. They had made the error of keeping their executor accounts and a Mexican bond of £30 15s. in Bull's room, conceiving the 'Journal Office of the House perfectly secure from danger by fire'. The bond belonged to the youngest daughter of Mr Gunnell, who was then left in straitened circumstances: her inheritance went up in flames.[32] Charles Pole, a committee clerk, had deposited 'for safety' in his office 'various property, in books, desks, jewelry etc to the value in the whole of about £35', which was destroyed.[33] Robert Collins, a junior clerk in the Vote Office, lost at least £10 in petty cash he had set aside for paying the porters to deliver official papers to MPs; presumably this was his own money and he intended that the House would later reimburse him. He also owned a nine-volume set of David Hume's *History of England*, and—beside 'many other things too numerous to mention'—five volumes on the *Elements of Rigging*. This was a bizarre reference book for a clerk to own.

As his career in the House is not otherwise recorded, perhaps he ran away to sea soon after.[34] Richard Jones, Assistant Clerk of Elections, and John Walmisley, another clerk, lost much in the fire. They were responsible for drafting bills for people who needed a local or private Act passed, outside the normal run of government Acts, and later set up in business together as Parliamentary agents. Jones lost two hundred volumes of private Acts 1767–1834 worth £210 and other bound books, treatises, and reports valued at £50. Walmisley lost his reports, an index to Acts, digests, books on parliamentary practice, the lease of his house 'kept at the House for security', two valuable cornelian seals, one set in gold with a family crest; and another cornelian seal, with his own crest, totalling nearly £82.[35] The less well-off staff of the House suffered even more. William Bevan, a messenger, had his uniform of dress hat, dress shoes, and dress buckles, worth £1 12s. 6d., destroyed.[36]

John Rickman, the Clerk Assistant of the Commons, had not left any property in his office.[37] Instead a greater misfortune beckoned. His whole house, along with all his possessions and those of his family, was now under threat. Aged 63 in 1834, he was by inclination and background a statistician. He had come to the Commons administration via a circuitous route. His work on population and demography had been shown to Speaker Abbot, and this had resulted in the 1801 Census Act, creating the first national statistical survey: an idea which Rickman had been promoting for some time, and which still takes place in the UK today every ten years. He was subsequently appointed secretary to Abbot, which allowed him to occupy a house in the Palace precincts (on the riverside opposite the Speaker's House) where he brought up his family. In 1820 he then became Clerk Assistant, which necessitated a move of residence round the corner, next door to Westminster Hall. He also had close friendships with a number of well-known literary figures, including the writer Charles Lamb and the poet Robert Southey.[38] Susannah Rickman, his sensible wife of nearly thirty years, was overcome by an unaccustomed fit of hysteria on realizing the danger they were now in, upon awaking from her evening nap. But

Frances Rickman, their level-headed and practical daughter, kept her nerve and immediately set about preparing the household to evacuate.[39] Her brother and sister were away from home. What had seemed like a small fire at the other end of the building thirty minutes previously was now rapidly gaining pace towards their house, blown to their north-eastern corner of the Palace precincts. Mrs Rickman meanwhile recovered herself and began to pack up speedily, 'as if for a long journey'.[40] Over the next seven hours, the family, their servants, and a host of friends engaged in a feverish race against time to save everything from the oncoming flames.

By half past seven, the fire had completely gutted the interior of the Lords Chamber, Howard's coffee-house, and all the western rooms on the principal floor, including the offices of the receiver of fees, the Earl Marshal, the Lords' Clerk Assistant, the Printed Paper office, and the room occupied by the Lord Chancellor's staff. A few minutes afterwards there was a tremendous roar, and part of the roof and ceilings of Wyatt's crenellated, jerry-built frontage containing those rooms fell in: 'and when the smoke and sparks that followed the crash of the heavy burning mass that fell had cleared away, nothing met the eye but an unsightly ruin, tinted with the dark red glare reflected from the smoldering embers at its feet'.[41] The entire roof of the western flank of the House of Lords then began falling in at intervals, 'each portion as it fell sending up immense volumes of flames which gave the surrounding neighbourhood the brilliancy of noonday'.[42]

Augustus Welby Northmore Pugin was delighted. In London for a few weeks following his aunt's death in early September, the up-and-coming architect and designer witnessed the fire from almost its start through to the end.[43] He had always considered the façade of the House of Lords and the newer buildings behind a hideous architectural mistake, and gleefully declared:

> there is nothing much to regret & a great deal to rejoice in. A vast quantity of Soanes mixtures and Wyatts heresies have been effectually consigned to oblivion. Oh it was a glorious sight to see his composition

mullions & cement pinncles [*sic*] & battlements flying and cracking
While his 2.6 turrets were smoaking Like so many manufactoring [*sic*]
chimnies till the heat shivered them into a thousand pieces—the old
walls stood triumphantly amidst this scene of ruin while brick walls &
frames sashes slate roofs &c fell faster than a pack of cards.[44]

Meanwhile, a Brighton coach was travelling towards London over the
top of the South Downs when the strange red light on the horizon
indicated to passengers that a great conflagration was under way.
Among them was another architect, Charles Barry. No sooner had the
coach reached his office, than he hurried to Westminster, and remained there all night. He found that 'all London was out, absorbed
in the grandeur and terror of the sight'. Barry had been born and
brought up in a house overlooking New Palace Yard, at 2 Bridge
Street. In 1812, at the age of 17, he began exhibiting his architectural
survey drawings at the Royal Academy, starting with an interior of
Westminster Hall. The disaster unfolding before his eyes triggered a
highly ambivalent response, and his son later recounted how 'the
thought of this great opportunity, and the conception of designs for
the future, mingled in Mr Barry's mind, as in the minds of many other
spectators, with those more obviously suggested by the spectacle
itself.'[45] This seems to be the source of the words Barry is supposed to
have uttered on realizing the old Palace was doomed: 'What a chance
for an architect!'[46]

Just about the same time as the Lords' façade fell in, James Braidwood arrived on site, an hour or so after the fire was first seen outside
the building. He began directing the operations of his force at the scene
immediately, placing them in the positions he felt were most efficient to
extinguish the flames.[47] Two fire engines and fourteen firemen had arrived from LFEE District A; another five engines with twenty-three men
from District B; District C supplied a further four engines plus sixteen
firemen; while two engines had come from Districts D and E with six
and five firemen respectively.[48] The thirteen horse-drawn engines were
manually operated, each with 260 feet (79m) of leather hose, made by

the firemen, and with twelve feet of suction. That meant that the engines had to be placed within twelve feet of a water source to operate—either from a street-plug, a natural water source, or connected to another engine's hoses to extend their reach. These constraints were the key to what the firemen were, and were not, able to do during the long night that followed. The uniforms of the sixty-four LFEE firemen distinguished them from the army and police and, most importantly, from the other gaudily dressed insurance company firemen present. Theirs was a single-breasted, dark grey frockcoat and trousers, black leather helmet and blue cap, with a red badge on the left breast sporting each man's number.[49] Braidwood first directed his attention to extinguishing the fire in the House of Lords where it had first broken out. It was hopeless. 'In consequence of the flames having got complete possession of these buildings and issueing [*sic*] through the roof previous to my arrival,' he wrote later, 'I found it impossible to prevent their destruction.' Braidwood decided to cut his losses and turned his attention to the threat to Westminster Hall and the Courts of Law in front of them.[50]

As early as seven o'clock, considerable alarm was already being felt in relation to the Great Hall. The 'great body of flames' and particularly the 'flakes of fire' which so many commentators mention in their accounts were being blown briskly over its roof by the wind.[51] Its southern gable end was little more than 30 feet (9m) from the northern end of the burning Lords frontage—with only a corridor and a single room on each floor of the Commons separating the two.[52] At a little after seven, people hurrying round to the main entrance of the Hall in New Palace Yard found that the strong, iron-railed gate in front was locked fast. Beyond the railings, its great north door was open.[53] Through it, Henry Yevele's huge gothic window, at the far end of the Hall, could be seen, lit up by flames 'burning most furiously against the splendid window at the south end of the wall and also on the eastern and western side'.[54] The Hall was deserted. The iron fence could not be broken or scaled. Then an impetuous band of gentlemen, 'who felt the nature of the emergency', returned to Old Palace Yard, and succeeded in persuading some firemen to batter in a small postern door at that end which led into the Hall. It took 'great exertion', but they finally burst through. There they found themselves in 'a short passage ten paces in length' between the great window and the windows on the wall oppo-site where the fire was breaking through.[55]

Following his arrival and assessment of the dire situation, Braidwood drew an imaginary line from the south end of Westminster Hall towards the river, intending that the fire should be confined to this boundary—anything north of it, including the Hall and the Speaker's House, had to be protected.[56] To effect this, he placed two engines at the large south window of the Hall, one of which he afterwards removed to the outside of the roof on the west side. Two were stationed in New Palace Yard to serve these because of the great distance from the water. (This implies that somehow the iron railings at the north door had now been unlocked or somehow cut through to allow the LFEE engines access.) Two were placed on the east side of the Hall outside the roof. One engine each played on the south end of the roof of the Courts of Law, inside the Courts of Law, and on a low roof between the Courts and the Hall. One engine was sent round to the Speaker's House. Three were moved to the southernmost end of the fire to cover the Library of the House of Lords whose treasures were in dire need of protection.[57]

The professional nature of the LFEE activities did not pass without comment. They were qualitatively different from the other firefighting attempts. 'During this time', wrote *The Times*, the day after,

> several engines arrived and were placed in Westminster Hall, so as to play with advantage on the Speaker's premises, and also on the back part of the Hall itself, which was not threatened with destruc-tion, the body of flames showing through the windows and com-pletely illuminating the Hall. Engines were also placed in front of both Houses in [St Margaret's Street], and then the pipes were like-wise brought over the tops of the houses in Abingdon-street to play upon the fire. No time or exertion was lost in obtaining the fire-ladders which are kept in the passage leading from Bridge-street to the Hall.[58]

In fact, the long fire-ladders, with their ingenious sliding construction, particularly excited the attention of spectators.[59]

Many fire engines continued to work outside Braidwood's control. It is unclear when most of them arrived; but as well as those which had driven up before seven o'clock, there were two more from the County and seven others at the scene from various other insurance companies including the British, West of England, and Norwich. It has also been suggested that there might have been a new-fangled steam fire engine in attendance which had the capacity to throw 150 gallons (680 litres) per minute to a height of 90 feet (27m) from a coke-based furnace, building up steam in twenty minutes.[60] It is true that there had been steam-powered engines under development from the late eighteenth century (which were less powerful but more portable than manual pumps, and of course required less manpower to work them), but in fact the LFEE did not take them up until 1852 as floating engines, or 1860 as land engines. This first steam fire engine, invented in 1828 by a Swede called Ericsson, apparently caused much envy in the LFEE but was later destroyed by a London mob. So if there was such an engine at the scene, it did not belong to the London Establishment.[61] Of the estimated twenty-eight or twenty-nine engines present at the fire during the night, thirteen belonged to the LFEE. The rest were independent and belonged to the dissenting insurance companies, the local parishes, and government buildings—independent not only of Braidwood, but also of each other. The recent session of Parliament had been considering bringing parochial engines under LFEE control, but none of the rest. 'The whole number ought to have been under the control of a single superintendent,' thundered a letter to *The Times* a week later, having seen the chaos that these fragmented arrangements brought about on that night.[62]

Around this time, Captain George Manby was struggling to cross Westminster Bridge. Some half an hour before, he had been finishing his dinner in his house in Lambeth when, hearing a yell of *Fire!* outside ('a cry that always calls the energies of my mind into action'), he had rushed to his window. He looked down to the end of Paris Street where he lived. There, between the rows of houses on either side was the river, and, across the river, 'one awful, raging, sheet of fire'. Abandoning his meal, he left the house and went, as everyone else did that night, to view the spectacle.[63] Manby watched the burning with an expert eye. He was well known as the inventor of various kinds of lifesaving equipment. His most famous contraption was a mortar-propelled rescue rope to establish a landline for shipwrecks known as the Manby Apparatus, and other kit to save people who had fallen through ice. But he had also, in 1816, demonstrated the first fire *extincteur*, made of a portable vessel holding a fire-extinguishing solution under pressure.[64] By the time he got to Westminster Bridge it was blocked by spectators, and he began to make the best haste he could through an 'almost impossible crowd'.[65]

The collapse of the roof of the west front of the House of Lords signalled the next phase in the progress of the disaster. The tumbling rafters had added fuel to the fire so that 'for the next half hour the majestic buildings presented the appearance of a burning city'. As eight o'clock approached the whole range south of Westminster Hall, from the portico of the Peers' Entrance at the southern end to the corner where it communicated with the House of Commons committee rooms, was ablaze. It was now penetrating from west to east too, towards the river, which unless something was done would take it directly through St Stephen's. From the river, the fire was particularly grand and impressive. The initial impression from the boats was that nothing could save Westminster Hall from the flames, and the blaze seemed to be developing a personality and will of its own:

> There was an immense pillar of bright clear fire springing up behind it, and a cloud of white, yet dazzling smoke, careering above it, through which, as it was parted by the wind, you could occasionally perceive the lantern and pinnacles, by which the building is ornamented. At the same time, a shower of fiery particles appeared to be falling upon it with such unceasing rapidity as to render it miraculous that the roof did not

burst out into one general blaze. Till you passed through Westminster Bridge, you could not catch a glimpse of the fire in detail—you had only before you the certainty that the fire was of greater magnitude than usual, but of its mischievous shape and its real extent you could form no conception.[66]

Confronted by the inferno, and with his options swiftly decreasing, James Braidwood sent for one of his few remaining pieces of equipment: the Establishment's great Floating Engine.[67]

7

Immense and appalling splendour

S HORTLY BEFORE EIGHT o'clock, the wind veered more westerly, pushing the flames right onto the House of Commons
buildings. The angle which abutted the House of Lords immediately caught fire and, despite the 'utmost exertions' of the firemen
and military, the roof of St Stephen's ignited, where, the woodwork
'being old and dry, the flames spread with the rapidity of wild fire'.[1]
The speed of the fire as it spread into the Commons 'was truly astonishing', wrote Pugin, 'from the time of the House of Commons first
taking fire till the flames were rushing out of every apperture [sic] it
could not be more than five or six minutes and the effect of the fire
behind the tracery &c was truly curious and awfully grand'.[2]

Almost exactly on the hour, the burning timbers gave way, and the
roof of St Stephen's fell in with a stupendous crash, accompanied by
'an immense volume of flame and smoke, and emitting in every direction millions of sparks and flakes of fire'.[3] Just minutes before, Duncannon, the Home Secretary, had been viewing the fire from its roof.

He was said to have gone up there with a party to watch and superintend the playing of water from the fire engines. The fast-approaching blaze left them in considerable danger, but it was reported that he refused to leave the roof until all the firemen and soldiers who were with him had first descended. Two minutes after they left, the red-hot timbers plunged to the ground.[4]

The enormous noise and sudden burst of flame were mistaken by some as 'resembling the report of a piece of heavy ordnance'. The belief that cannon were being let off led to panic among the assembled crowd. Others thought that the explosion was due to a magazine of gunpowder igniting, which was then expected to blow up further at any moment. Many bystanders were seen making desperate efforts to escape the scene.[5] The confusion which followed baffled the power of description, according to one newspaper:

> in the moment of terror and dismay occasioned by the apprehension of other and perhaps more mischievous explosions, a report having gone abroad that large quantities of powder and other combustible materials were deposited under the House of Commons, no one thought of anything but consulting his own safety by flight and a general retreat took place, in the course of which various accidents occurred extending in some instances even to fracture of limbs; but so far as we could ascertain on the instant, none threatening to be attended with fatal consequences.[6]

It was probably this incident which inspired some broadsheet doggerel, published a few days later:

> Thousands flocked in every direction,
> The engines flew to give protection,
> The father, daughter, son and mother,
> Was all a tumbling o'er each other,
> The poor old Abbey with fear was quaking
> And Westminster Hall was sadly shaking,
> Some cried the country wil [*sic*] be undone,
> Some said Guy Faux was come to London.[7]

No doubt thoughts of the Gunpowder Plot sprang to everyone's mind, it being only three weeks before Bonfire Night. Suspicion of Catholics in ultra-Tory quarters and elsewhere had not died down simply because of the passing of the Catholic Relief Act. The verse went on:

> We never once thought such a rapture,
> Would seize that ancient Gothic structure
> Some thousands vowed it was amasing
> To see both Parliament houses blazing;
> There was thousands in great consternation
> And all were in deep conversation,
> For many years they braved the weather,
> But alas! they fell and died together.
>
> Some said what will the lords all do, sir—
> And all the Members of Parliament too sir,
> There will be struct with wonder dumb sir.
> For there [*sic*] seats are gone to Kingdom come sir,
> The wool-sack, throne and lots of treasure
> Was massacred in a dreadful measure.
> And it is not yet exactly stated
> How this dreadful fire originated.[8]

And now the realization dawned that, at St Stephen's, surrounded by flames on its west and southern sides, with the bulk of Westminster Hall and its own cloisters to the north, and without a street-plug nearby on the garden side, 'no engines could be brought near to afford any means for even attempting to check the progress of the fire; and the building having wood around it to cover the beautifully painted walls, and a wooden under-roof, was the aptest fuel for the flames of any where found'.[9] The only approach that could be made by rescuers was from the Thames. The shoreline was just fifteen or twenty yards from its east end.[10] It was an ideal opportunity for the LFEE's top-of-the-range piece of kit: the great Floating Engine. The pre-1833 fire insurance companies frequently used watermen plying up and down the Thames as firemen, as so many insured properties were either warehouses or riverside mansions, and this enabled a swift response in emergencies. So when it joined the LFEE, the Sun Fire Office had brought with it a large barge-mounted pump, which could shoot an endless supply of water from the safety of the river. It was the ideal thing for tackling a riverside blaze, and was ready to be called at any moment from its mooring at King's Stairs, Rotherhithe, beyond the Tower of London.[11] There was just one problem. As the Thames was approaching the lowest point of its ebb tide, the barge could not make its way the three and a half miles upriver to respond to Braidwood's summons until the water rose again. It was to be another six hours before 'the Float' arrived.

As a result, the work of destruction was 'sooner over than in the other House', noted *The Times* correspondent, grimly, and put it down to the greater amount of timber which the fabric of the Commons contained. The low tide also meant a very scanty supply of water from the shore was available to the one or two fire engines—'not very advantageously placed'—which had finally got going on the river side, as they attempted in vain to save that 'interesting edifice' from absolute destruction.[12] After the roof collapse, great columns of fire were seen to issue through the three large windows at the Chapel's east end. It was so completely engulfed that 'less than an hour after it had taken fire...nothing but the side walls and the two Gothic towers were left standing'.[13] Beneath the House of Commons Chamber was the original ground-floor undercroft of St Stephen's. By 1834, this had become the Speaker's state dining room, 'a curious, antique and magnificent chamber, singularly carved and decorated', which was believed to have been entirely destroyed in the collapse and subsequent blaze of the old Chapel above.[14]

Now, devouring the House of Commons Chamber, the fire continued to push north-eastwards towards the Speaker's House on the river frontage.[15] In the angle between St Stephen's and the Speaker's was the original double-tiered cloister of the collegiate foundation, complete with a miniature chapter house dating from the 1520s. Buried deep within the Palace complex, unseen by outsiders, this was one of the most beautiful structures at Westminster, 'the roof of which is scarcely surpassed by the exquisite beauty and richness of Henry the Seventh's chapel in the neighbouring Abbey', wrote a contemporary guidebook.[16] By 1834 it was part of the grand apartments in the Speaker's House. As the Commons Chamber blazed away, its roof caught alight, and the flames began to spread diagonally above its Perpendicular stone vault.

Dispatches were sent throughout the country to convey the terrible news, the one winging its way to the King reaching him within two hours of the fire beginning.[17] There was little need for him to be informed. William and Adelaide could see the glow in the night sky from Windsor, twenty miles away.[18] To those watching, the inferno in the Commons buildings now appeared to 'retrograde as well as advance'; that is, it spread to the rooms and buildings on the Abbey side, opposite Henry VII's Chapel, as well as towards the river. Although several fire engines were shifted around, they made no difference. Three fatigue parties belonging to the Guards arrived which 'assisted the firemen materially in their exertions and the working parties had sufficient scope for their labour at the engines owing to the excellent arrangements made by the Commissioners of Police for keeping off the pressure of the crowds assembled in every quarter'.[19] The danger to the Hall, however, appeared to be more imminent than ever. The range of committee rooms, situated immediately over the Members' Entrance to the House of Commons (in the wing known as the Stone Building, next to the Law Courts and opposite the Abbey), 'appeared to be entirely enveloped by the devouring element'. A dense black column of smoke issued from that part of the building, almost immediately followed by a large column of flame, surrounding the south end of the Hall with a semicircle of tongues of fire which were beginning to lick its gable end. At this point several engines were seen being wheeled into it, and 'an immense quantity of water was distributed over every part of the building'. The firemen and soldiers employed on the exterior redoubled their efforts, apparently wholly regardless of the danger to which they were exposed by the falling of burning rafters and the showers of molten lead which poured down on every side. The conflagration ultimately extended all round the 'new front buildings' of the Lords, utterly consuming the apartments of the Lord Chancellor, Mr Courtenay, and other offices ranging round to Howard's coffee-house, which was itself completely destroyed.[20] *The Times* described the ghastly scene at this moment. 'The wind had increased', it informed its readers,

> and as the flames were rapidly spreading and danger was apprehended of the Westminster Hall would take fire, three engines were sent round to the north entrance and the hose of each having been laid along the interior of the pavement the water was directed to that part of the roof of the Court of Exchequer. During the whole of this time, the fire appeared to be unabated, and immense volumes of fire, sparks, and smoke poured out of the various parts of the buildings, and were carried away in the direction of Westminster-bridge over the small houses called Cotton Garden and adjoining Westminster Hall. The scene of the conflagration at this period was awful in the extreme. The light reflected from the flakes of fire, as it shone on the Abbey and the buildings in the vicinity had a most extraordinary effect, and every place in the neighbourhood was visible, so that a person would be enabled to read as in the daytime.[21]

The rapidity with which the Commons buildings were falling prey to the flames led some rescuers to risk all in saving some of the most precious items in the Palace. Mr A. H. Butt, the Deputy Serjeant at Arms, went after what was described as 'the Lord Chancellor's mace' (erroneously said by the newspapers to have been carried before Charles I at his execution) when the room where it was stored caught fire. He placed a ladder at the window and two firemen, Mr Birch and Mr Hill, 'gallantly mounted' to the second floor. Having smashed the window with their axes, they broke open the cupboard where it was deposited, and handed the £400 silver-gilt mace, dating from the restoration of Charles II, out to Mr Butt, who then took it home for safety to his house in Abingdon Street.[22] Afterwards, a proposed reward of £2 2s. for the firemen was disallowed on account of them just doing their

PLATE 1   Old Palace Yard in 1834, from the south-west. In the centre, Wyatt's 'cotton mill' frontage to the House of Lords, with Soane's King's Entrance at the far right. On the left, the House of Commons wing housing the Members' Entrance, committee rooms, and Bellamy's, with the southern gable of Westminster Hall behind.

PLATE 2   New Palace Yard in 1834, from the north-west, showing the north door of Westminster Hall with Soane's neoclassical Law Courts on its western flank. The building to the left of the north door was the Rickmans' house, with the gabled Exchequer buildings beyond, on the east side of the Yard. On the far right, the east end of Westminster Abbey.

PLATE 3 Interior of the House of Lords, with the royal throne at the south end, and the Armada Tapestries on the walls. Black Rod's Box, where the fire first emerged, can be seen in the curtained area bottom right.

PLATE 4 The stuffy, hot, and cramped panelled interior of the House of Commons where MPs were packed in 'like herrings in a barrel'.

PLATE 5 Cross-section of the House of Commons, drawn in August 1834, from west to east. Note the uneven floor levels and multiple partitions, as well as the way on the right the original chapel of St Stephen's has been divided horizontally by Wren. Women could only view proceedings in the Chamber, unofficially, through the ventilation shaft in the roof of the chapel. To the left of the window of Westminster Hall, a cramped and unceremonious staircase leads from the Members' Entrance to the principal floor, and on the far left there are ground-floor waiting rooms and first-floor committee rooms, with Bellamy's dining rooms on the second floor, and the Bellamy family's apartment in the attic. The lead roof or 'flat' where the LFEE firemen struggled to contain the blaze from reaching Westminster Hall can be seen in the very centre, at the base of the great south window.

PLATE 6 Thirteenth-century tallies. The tally-sticks burnt in 1834 dated from the 1820s or slightly earlier, but would have looked very similar.

PLATE 7. One of the Armada Tapestries, as engraved by John Pine in 1739. This shows the engagement of De Valdez's galleon by Drake off Plymouth, complete with gambolling dolphins. The original tapestry this depicts was probably the one which survived the fire.

PLATE 8 Private insurance company horse-drawn fire engines speed towards a fire around 1830. This engraving gives an idea of the excitement and spectacle which even an ordinary fire generated.

PLATE 9
James Braidwood (1800–61),
Chief Superintendent of the
London Fire Engine Establish-
ment, photographed in later life.

PLATE 10   A 20-man, horse-drawn, manually pumped London Fire Brigade Engine of the
1860s. The engines in use by Braidwood and the LFEE in 1834 were probably around two-
thirds this size. Firemen would have ridden on the tank on the way to the fire.

PLATE 11 Chance at the 1834 fire, one of three engravings of this famous canine by William Heath. His collar reads: 'Stop me not, but onward let me jog | For I am Chance, the London firemen's dog!'

PLATE 12 The fire probably towards 8 p.m., in a crude but atmospheric engraving, looking northwards across Old Palace Yard. The space has been cleared by the police and guardsmen, and the parish fire engines are in attendance as the House of Lords frontage burns furiously and the House of Commons committee rooms begin smoking.

PLATE 13 William Heath's view of the fire slightly later from a similar position, 'from a sketch taken by him by the light of the flames at the end of Abingdon Street'. People lean out of windows in nearby houses to get a better view, and a cart removes books and papers from the site. St Margaret's Church can be seen in the background, beyond the Henry VII Chapel at the east end of Westminster Abbey. Chance is a tiny figure in the centre, just beneath a great billow of white smoke.

PLATE 14 J. M. W. Turner's watercolour of the fire from the same position in Old Palace Yard, later still, with the massed crowds and a gushing street-plug in the foreground.

PLATE 15 Turner's 'golden apocalyptic glory' based on watercolour sketches made at the scene or shortly afterwards. This is the view from the Lambeth side of Westminster Bridge.

PLATE 16 Turner's other oil painting of *The Burning of the Houses of Lords and Commons* from further downriver near Waterloo Bridge, showing the Floating Engine steaming towards the scene in the lower right corner, as the night sky over London blazes.

job.[23] But the Deputy Serjeant's son Mr John Butt, a solicitor of College Street, who also 'rendered most essential services' (perhaps during this exploit), was injured, and his surgeon's bill of £10 10s. was paid in compensation.[24]

Two account books from the Serjeant at Arms' office were saved, possibly grabbed by the firemen as they left with the mace, though that cannot be proved. They survive today in the Parliamentary Archives, one with heavily singed pages, the other with its leather covers blistered from the heat. These rare survivals may equally well have been thrown out of a window elsewhere.[25] A number of soldiers from the detachments of the Horse and Foot Guards were employed in getting out the property in the remainder of the House of Commons, the destruction of which now appeared inevitable.[26] 'Every exertion was made', wrote the *Gentleman's Magazine*, 'to save the public papers and other important documents, vast quantities of which were conveyed to a place of safety, although many were unfortunately consumed.'[27] The clerks and other officials of the House, with the help of the police and military, continued to requisition whatever transport they could find, and loaded up immense quantities of furniture, books, and papers, both loose and boxed, which were then taken to various houses in the neighbourhood of Palace Yard, often the clerks' own houses. Scores of loads were moved: in carts, wagons, hackney coaches, cabs, 'chariots', and vehicles of every description. The greater part of the books and records were now being deposited across the road in St Margaret's Church, which had been opened up under the direction of Mr Schofield, one of the Marlborough Street officers; salvaged furniture was placed outside in the graveyard.[28] By half past eight, this area was starting to fill up. Efforts to save the official records soon became more desperate. As the fire raged at the southern end of the building, people were seen throwing books and records of various descriptions out of the windows and into the street. Parties stationed below were catching them and sending them to safety in the church.[29]

107

St Margaret's Churchyard was not regarded as an entirely respectable place in 1834, though it was good enough as a furniture store in an
emergency. It was in fact a well-known cruising ground (like other
London parks) and only the year before, the MP William Bankes had—
notoriously—been arrested for cottaging in a public convenience there.
Following humiliating questioning by the police, involving inspection
of his underclothes, Bankes was charged with 'attempting to commit
an unnatural offence' with a soldier, Private Flowers. Bankes claimed
that he had been caught short on his way to the House following a
dinner with Lord Liverpool; an excuse which was viewed with derision
by some, given that he lived almost next door at 5 Old Palace Yard. At
the time, sodomy was a capital offence, and Bankes had to be protected
from a baying mob of 2,000 people when leaving the police station in
Tothill Street, having been bailed. Bankes was acquitted six months
later (the Duke of Wellington, among others, proving an eloquent
character witness), but his story touches on the 1834 fire in several other
ways too.[30] His family controlled the constituency of Corfe Castle—
one of the rottenest of rotten boroughs—and not surprisingly he was a
staunch anti-Reformer, as well being against Catholic Relief. William's
father, Henry Bankes, was one of the MPs so critical about excessive
expenditure on Buckingham Palace in the years before the fire. In 1833
the death penalty, with which this arch-conservative was threatened,
had just undergone significant reform, like so much else. Stealing
horses, cattle, and letters; forgery, smuggling, and slave trading; and
certain types of attempted burglary, robbery, piracy, and murder; plus
offences against the Riot Act, were no longer punishable by hanging.
But open homosexuality still was, even though the term had not yet
been coined. When Parliament was dissolved by the King in November
1834, Bankes retired as an MP, though he remained acquainted with
Lord Broughton, whom he had known at Cambridge and with whom
he shared a mutual friend in Byron. Then, in 1835 he engaged Charles
Barry to redesign the family seat at Kingston Lacy in Dorset, whom he
drove to distraction with his demands; and in June 1841 (as a well-

known connoisseur and man of extraordinary taste) he was invited to submit his views to the committee considering artworks for Barry's new Palace. Sadly, two months later, he was once again caught in a compromising situation with a guardsman, this time in Green Park, and fled the country for good, spending the rest of his life in self-imposed exile in Venice.[31]

About half past eight o'clock the fire arrived at such a height that all the Commons' offices in the Stone building were burning with 'the greatest fury'. Many fire hoses continued to be directed to this quarter to prevent, if possible, the flames from communicating with the Courts of Law and Westminster Hall.[32] The blaze had been raging for two hours. Those who watched—and there were tens of thousands of them—never forgot the experience. It was an extraordinary sight. People who found themselves on Westminster Bridge noted how the houses opposite the Palace were as if 'illumined for some great occasion', while all the time being battered by 'very annoying' sparks as they spectated.[33] Streets and bridges were now hopelessly clogged with people and vehicles, but there was still space on the river for enterprising onlookers to get a closer look, where 'innumerable wherries, which were plying during the whole time, [were] taking persons abreast of the House of Commons'.[34] Ralph Wornum the art critic (and later keeper of the National Gallery) was in Munich at the time, but a friend sent him a copy of the *Weekly Dispatch*, describing the disaster as

the grandest & most superb one I remember—I was present for four hours on *every* side of it—and took a boat and rowing by the side of the fire saw the effect of its magnificence to the greatest advantage—I thought of you and much wished for the ability of taking a sketch of it—for the mixture of flame and smoke, of volumes of sparks rising to the skies, of a full moon & starlight night—the reflection in the water, the crowds of persons in all directions (on the tops of the houses, on the Abbey leads, lining the bridges and water side) and pouring in on all sides even from Deptford—and above all the Abbey itself being lighted up by the fire—the military—firemen and a thousand more concurrent circumstances all tended to inspire one's feelings with an impressiveness I shall ever remember—it was indeed most horridly beautiful...[35]

The now-massive blaze spectacularly floodlit Westminster Abbey, especially the towers at its west end. The crowds gaped in amazement. Its 'architectural beauties were never seen to greater advantage than when lighted by the flames of this unfortunate fire', wrote one spectator.[36] 'The different shades of colour assumed by the devouring element in its various evolutions presented a sight at once splendid, grand and the effect produced by the reflection of light and shade upon the various abutments and recesses of the venerable Abbey, and on the highly finished embellishments of that elegant structure, King Henry the Seventh's Chapel, formed a spectacle possessing so much of the sublime and beautiful that no language can convey an adequate description of its magnificence and grandeur', wrote another.[37] It was 'a perfect fairy scene'.[38]

Today, we would describe the scene as cinematic. The intense illumination provided by the flames on their surroundings was something which people used only to candles and oil-lamps at night found astonishing and quite outside their everyday experience. At a time when even limelight in the theatre was unknown (first used in public at Covent Garden in 1837), it was not surprising the enormity of the display got such a response. Occasionally Londoners might experience illuminations on state occasions, and fireworks were a regular occurrence in the pleasure gardens at Vauxhall.[39] Technologies of light—the 'new media' of the early nineteenth century, as they have been described—such as kaleidoscopes (invented in 1816) and magic lanterns were extremely popular.[40] But the nearest thing in people's contemporary experience to a cinematic spectacle would have been the diorama. This was a ten- or fifteen-minute show during which the audience would gaze in darkness down an opaque canvas tunnel to a scene (or scenes) painted on double-sided transparent canvas, which would appear to move or transform, depending on changes in the lighting, the use of coloured silk curtains, and movable skylight blinds.[41] Standing on a turntable, the audience would then be whirled round to see a different set of scenes. It was like a very slow cartoon with only a few frames. In 1831, Adam Lee, as a sideline to his main job as Clerk of Works, had exhibited a series of 'Cosmoramic and Dioramic Delineations' of the Palace of Westminster at the Society of Watercolour Painters in Pall Mall. ('Cosmorama' indicates that the pictures were viewed through mirrors and lenses, and specially illuminated to give the appearance of reality.) The aim of the exhibition (entrance one shilling) was to generate enough subscribers to publish a volume of such views (priced at six guineas plain, ten guineas coloured).[42] But not even the great Heptaplasiesoptron set up in Vauxhall Gardens in 1822—an illuminated water feature with palm trees, serpents, mirrors, draperies, and the 'seven times reflected effulgence' of its variegated lamps—would have prepared anyone for the evening of 16 October 1834.[43]

The packed crowds accordingly became an audience, with an audience's reactions, so that 'whenever a more than usually brilliant body of flame burst forth, and whenever the sparks and burning embers fell thickest among them, a simultaneous clapping of hands took place, as if the scene were a mere theatrical exhibition'.[44] Indeed, the crowd exhibited the kind of suspension of disbelief any rapt audience would, when sitting behind the invisible fourth wall of a stage production. They applauded 'as though they had been present at the closing scene of some dramatic spectacle, when all that the pencil and pyrotechnic skill can effect is put into action, to produce a striking coup d'oeil'.[45] Westminster Bridge was the dress circle overlooking the stage, the sight of which fired the imaginations of the many artists and writers present. Letitia Landon, the celebrity poetess and novelist (and influence on Christina Rossetti and Elizabeth Barrett Browning), declared,

> Never was a spectacle so much enjoyed. All London went to see the fire—and a very beautiful fire it was. One most singular effect was produced by the mass of people on Westminster Bridge as seen from the river. Between the white pillars of the ballustrade [*sic*] was a human head and the contrast of light and dark, was as they say of a scene in a new play, very effective. The sky too had a lovely appearance—on one side loaded with crimson clouds, on the other the moon breaking through the light fleecy vapours—with a space of such clear blue tranquility [*sic*] around.[46]

Large sums of money were paid by the nobility and gentry for seats at the front windows of the houses in Bridge Street and Parliament Street, 'to witness the imposing scene on the night'.[47] If these were the balcony boxes, then the river itself was the stalls. A *Times* correspondent in a boat watched as 'the dark masses of individuals formed a striking contrast with the clean white stone of which it is built, and which stood out well and boldly in the clear moonlight. As you approached the bridge you caught a sight through its arches of a motley multitude assembled on the strand below the Speaker's garden, and gazing with intense eagerness on the progress of the flames. Above them were the seen the dark caps of the Fusilier Guards, who were stationed in the garden itself to prevent the approach of unwelcome intruders.'[48]

The artist Benjamin Haydon (1786–1846)—flamboyant, narcissistic, and perennially broke—had had an audience with Lord Grey earlier that year. At it he had boldly proposed the replacement of the Armada Tapestries with a new cycle of his own historical paintings, but the Prime Minister had merely replied smoothly, 'They are a fine series, but there is no intention I know of to take down the tapestry, and the House of Commons is in such a temper about expenditure, that I could not propose such a thing.' Never one to hide his light under a bushel, Haydon realized on the night of the fire that it presented him with an unparalleled opportunity to resurrect his grand scheme. 'Good God!' he wrote in his journal that night,

> I am just returned from the terrific burning of the Houses of Parliament. Mary [his sister] and I went in a cab, and drove over the bridge. From the bridge it was sublime. We alighted, and went into a room of a public-house, which was full. The feeling among the people was extraordinary—jokes and radicalism universal. If Ministers had heard the shrewd sense and intelligence of these drunken remarks! I hurried Mary away. Good God, and are that throne and tapestry gone with all their associations! The comfort is there is now a better prospect of painting a House of Lords. Lord Grey said there was no intention of taking the tapestry down—little did he think how soon it would go.

Yet on proposing his scheme once more, this time to Melbourne, he was to be disappointed again, and slipped further and further into penury, leading ultimately to his suicide some twelve years later.[49]

To capture the once-in-a-lifetime scene at Westminster, many other artists besides Haydon flocked to the Palace, just as they did to sketch the smoking ruins in the days and weeks afterwards. At the Royal Academy on the Strand, the students climbed onto the roof of their college at Somerset House to view the disaster better round the bend in the Thames, having been alerted by the Academy porter in the library: 'Now, gentlemen; now you young architects, there's a fine chance for you; the Parliament house is all afire!'[50] Yet more went down to the river and other places to get closer to sketch. All agreed that the view from the Thames was one of 'immense and appalling splendour'.[51] But it also encouraged the foolhardy to take to the water. Two young men called Harland and Wybrow, who had hired a boat but were inexperienced in rowing, ran against one of the piers of Westminster Bridge, capsized, and would have drowned had they not been saved by a waterman called Pridham.[52] It was a good night for the watermen's profits, in general.[53] One of the art students, John Green Waller, missed the fire, and regretfully wrote in his diary the morning after,

> they all describe it as being the most grand and imposing sight they ever saw. The appearance of the Abbey lighted up by flames, they say, was most splendid and lately a scene of the most terrific grandeur ... The night was for the most part bright moonlight, which added greatly to the variety of effect, and the tide was ebbing at this time and this rippling made the grand picture still more perfect. I heard many say the scene from Waterloo Bridge as being the sublimest picture they ever witnessed. The glowing accounts I received from the students made me much regret I was absent from the Academy last night.[54]

In one of the other boats on that rippling water were the artists J. M. W. Turner and Clarkson Stanfield (also a marine and landscape painter), alongside some of the Academy students.[55] 'I never lose an accident,'[56] confessed Turner, and that night he witnessed the most climactic accident of his whole life. He spent until daybreak on the Thames or among the crowd on Westminster Bridge, the Lambeth side of the river, or at the southern end of the Palace in Abingdon Street, seeking the best vantage point—as was his custom when assessing a subject. From that night, Turner produced two sketchbooks of work: one containing nine watercolour studies and the other some minimal pencil drawings, although there is a dispute among scholars over whether he sketched them on the spot or immediately after the fire. The pencil drawings, taken down in a small pocket-sized notebook with marbled paper covers, were declared 'worthless' by the critic John Ruskin, before it was understood what the few strokes depicted. The watercolours were contained in a larger, hardcover, sketch block with green marbled covers.[57] In an elegant coincidence, these notebooks are now at Tate Britain, itself built on part of the site occupied by the Millbank Penitentiary where Joshua Cross had done time a few years before the fire. The great artist took down only what impressions he felt would spark his memory later in order to compose two of his most celebrated oil paintings, *Burning of the Houses of Parliament*, fascinated with the elemental forces of nature—earth, water, air, and fire—and watching their deconstruction before his very eyes on that autumn night into what one art historian has called a 'golden apocalyptic glory'.[58]

Meanwhile, another great landscape painter—John Constable— was sitting comfortably in a hackney carriage with his two eldest sons, parked on Westminster Bridge. Small watercolours of the fire and afterwards by him also survive. He later sketched the blaze for a friend as an after-dinner entertainment, who recalled,

> While describing the fire, he drew with a pen, on half a sheet of letter paper, Westminster Hall, as it showed itself during the conflagration; blotting the light and shade with ink, which he rubbed with his finger where he wished it to be lightest. He then, on another half sheet, added the towers of the Abbey and that of St Margaret's Church, and the papers, being joined, form a very grand sketch of the whole scene.[59]

George Manby, the fire extinguisher inventor, had now pushed his way over Westminster Bridge to the front of the Houses of Parliament, and, deciding there was nothing he could do to help, took refuge by the Canning statue. The Abbey, he thought, was made so brilliant by the flames that to his mind, 'it would be impossible for the hand of the most celebrated artist to imitate, or the most fertile imagination to conceive'.[60] Perhaps he was not too far away from the sketching Turner. The men had another connection as well. In 1831, Turner had exhibited his new painting, *Life-boat and Manby apparatus going off to a Stranded Vessel making Signal (Blue Lights) of Distress*, at the Royal Academy: a tribute to Manby based on sketches of a foundering ship he had seen at Great Yarmouth in 1824.[61]

As well as the visual impact the fire made on spectators, adding to the fearful spectacle was the incredible cacophony associated with the rescue attempts. There was the roaring of the flames; the 'dire yell'; the tolling of the bell of St Margaret's; the firemen shouting; the heated window-glass shattering (which 'made a report similar to that from firearms'); the burning timbers cracking; the scattering of tiles and slates; the crash of falling roofs and beams; the tramp of marching soldiers; the rush of water through the hoses; the drumming of the Foot Guards beating to arms; the rattling and thundering as the engines drew up on the pavement; the horse clarions wailing; and the occasional scream of some poor wretch whose foot had been run over by a wheel. It was, by all accounts, 'an awful accompaniment'.[62] The wind moaned round the walls and buttresses, 'hymning a dirge to expiring greatness and departing glories'. There were regular cries of 'Make room for my Lord A!' or 'The Right Hon, Mr B, this way!',[63] while Braidwood whistled instructions to his firefighters using coded signals on a high-pitched bosun's pipe.[64] And perhaps through the noise it was also possible to hear Chance, the firemen's celebrity mascot, barking in excitement.

Chance, a good-looking tan terrier of indeterminate parentage, attracted attention and myth in equal measure. He was said originally to have been a weaver's dog from Spitalfields, and first became acquainted with firemen at a prolonged fire in the East End. Apparently, the dog particularly liked the grey uniforms of the LFEE men, and soon made their Watling Street headquarters his home. It was alleged that once his owner saw that Chance preferred it there, he gave the twelve-year-old mutt to the LFEE along with a brand new brass collar engraved with the couplet:

> Stop me not, but onward let me jog
> For I am Chance the London firemen's dog!

From then on, the terrier regularly attended fires with the engines, barking in delight at the scene, supposedly walking along burning joists and timbers during the fiercest of blazes, and 'enjoying the luxury of a bath in the jet from the water main'. It was said that he got badly burnt on occasion, kicked in the jaw by a cruel passer-by at a fire in the Barbican on another, and even had his front legs crushed when run over by a fire engine.[65] If this is so, he made a series of miraculous recoveries, but he was certainly a real dog with an engraved collar and well-known enough to have his portrait painted several times. In one of the most evocative engravings of the fire—William Heath's lithograph of the west front of the Palace 'drawn by the light of the flames' at the head of Abingdon Street—smoke and flames billow out of the Lords and Commons, the huge crowds are transfixed in awe, people hang out of neighbouring buildings in Old Palace Yard to watch, and—at the centre of it all—the tiny figure of Chance barks encouragement to the firemen while the conflagration rages all around.[66]

Charles Dickens knew Chance too. Parliament 'does so little and talks so much' was his view for most of his life, and the dog's presence at the fire provided an unmissable opportunity for the burgeoning Parliamentary sketchwriter to poke fun at the vanity of one MP:

> He, and the celebrated fireman's dog, were observed to be remarkably active at the conflagration of the two Houses of Parliament—they both ran up and down, and in and out, getting under people's feet, and into everybody's way, fully impressed with the belief that they

were doing a great deal of good, and barking tremendously. The dog went quietly back to his kennel with the engine, but the gentleman kept up such an incessant noise for some weeks after the occurrence, that he became a positive nuisance. As no more parliamentary fires have occurred, however, and as he has consequently had no more opportunities of writing to the newspapers to relate how, by way of preserving pictures he cut them out of their frames, and performed other great national services, he has gradually relapsed into his old state of calmness.[67]

The MP in question was Mr W. Hughes Hughes, member for Oxford City, whose exploits (according to himself) included the clearance of the books in the Speaker's House 'in the most orderly and praiseworthy manner'. With the police, he then emptied some cabinets of private papers and jewellery, wrapped them in two large sheets, and had the bundles sent round for safekeeping to his home in Manchester Street, around eight o'clock. He then directed the 'careful removal' of portraits of past Speakers from the Levée Rooms, taking them out of their frames where necessary so they could be got through doorways, unhanging wall tapestries, and overseeing the removal of a magnificent chimney piece by carpenters from the Office of Woods and Forests. Nothing, he recounted to *The Times* at length, 'could be executed more satisfactorily'.[68] The reality was rather different, as the Speaker later discovered.

Incredible though it now seems, Dickens seems to have left behind no surviving correspondence on, or a direct narrative of, the fire. He was living at 18 Bentinck Street (off Cavendish Square in the West End) when the fire broke out but there is no known contemporary letter of his which mentions it.[69] In 1834 he was at the beginning of his writing career, and was already meeting with some success. Some years before he had taught himself shorthand, which enabled him to become a newspaper reporter first on legal cases and subsequently as a Parliamentary reporter for a new evening paper, *The True Sun*, in 1832. During 1834 he had been publishing short pieces on London life in the *Monthly*

Magazine using his pen-name 'Boz', and in August that year had been appointed as a reporter by the leading Whig paper, the *Morning Chronicle*, where he reported on provincial elections and other special events at a salary of five guineas a week. One month before the fire he had just taken on a commission by the *Morning Chronicle* to write a series of *Street Sketches*, also about London life.[70] It would be tempting to assume that some of the copy on the fire in the *Morning Chronicle* was written by Dickens, but this cannot be confirmed, and a listing of his known journalism does not suggest it.[71] He was a cub reporter at the time, and it would seem odd for him to get a chance to write about the biggest story of the century. It would, however, explain why he felt he had nothing more to say on the subject at the time.

However, lack of direct evidence does not mean Dickens was not there or was not influenced by this once-in-a-lifetime event which filled everyone's thoughts and conversation for weeks and months afterwards. Consider the following passage from *Oliver Twist*, first published in serial form in 1837, a book originally conceived as a protest against the effects of the new Poor Law. It is strangely similar to the accounts of known eyewitnesses, particularly that of Pugin. The thug Bill Sikes comes across a great blaze in progress as he flees the City after battering Nancy to death:

> Suddenly there arose upon the night-wind the noise of distant shouting, and the roar of voices mingled in alarm and wonder... The broad sky seemed on fire. Rising into the air with showers of sparks, and rolling one above the other, were sheets of flame, lighting the atmosphere for miles round, and driving clouds of smoke in the direction where he stood. The shouts grew louder as new voices swelled the roar, and he could hear the cry of Fire! mingled with the ringing of an alarm-bell, the fall of heavy bodies, and the crackling of flames as they twined round some new obstacle, and shot aloft as though refreshed by food. The noise increased... There were people there—men and women— light, bustle... others coming laden from the burning pile, amidst a shower of falling sparks, and the tumbling down of red-hot beams. The apertures, where doors and windows stood an hour ago, disclosed a mass of raging fire; walls rocked and crumbled into the burning well;

the molten lead and iron poured down, white hot, upon the ground. Women and children shrieked, and men encouraged each other with noisy shouts and cheers. The clanking of the engine-pumps, and the spirting and hissing of the water as it fell upon the blazing wood, added to the tremendous roar. Hither and thither he dived that night: now working at the pumps, and now hurrying through the smoke and flame, but never ceasing to engage himself . . . till morning dawned again, and only smoke and blackened ruins remained.[72]

Thereafter, the power of flame became a major theme in many of Dickens's novels, most notably in *Great Expectations* and *Bleak House*. It is more than tempting to assign at least some of his interest in the cathartic and dynamic nature of fire to events on that moonlit October night.

8

Damn the House of Commons!

THE FIRE NOW seemed to be everywhere. The entire flank of apartments overlooking Palace Yard was burning, including the two towers at either end. Soon after,

the central or projecting part of the building, where the stair into the Lords entrance to the throne end of the House [is] fell with a tremendous crash. The smoke afterwards was so dense as to obscure the flames for many minutes and to darken the atmosphere, but when they burst forth again, it was with terrific splendour increased. The octagonal tower facing Abingdon Street over the Peers' Entrance, the lower room of which was the Lord Chancellor's retiring or robing room, was a remarkable object during the progress of the conflagration. The different coloured flames which rushed from its windows and its roof amid so much ruin excited 'the most painful astonishment and admiration'.[1]

Presently, the upper portions of the Speaker's House were seen to be kindling by degrees.[2] The floors above the Members' Entrance to the Commons were raging furiously, and during the hour a large engine from the Royal Horse Guards barracks at Knightsbridge drew up, accompanied by a cavalry party of the 'Blues'. This set to work helping out another powerful engine from Elliot's Brewery at Pimlico (which had arrived early on the spot) in tackling that area of the blaze.[3] Reynolds, the Lords' firelighter, returned to the Palace to go on duty for his night shift. He had been asleep since early afternoon, and only found out about the disaster on his arrival at the scene. He was terror-struck and immediately thought to himself 'It was them sticks burning!' But he only said this aloud the following day once all the crowds had dispersed.[4]

The fire not only attacked the Palace across its west front, and back towards the river, but it had also pushed through the range of buildings leading to the Commons' Entrance in St Margaret's Street, facing the church. In that wing, adjoining the Law Courts, there were MPs' waiting rooms on the ground floor, committee rooms 11 and 12 above that, and above them, on the second floor, more committee rooms, along with Bellamy's kitchen and dining rooms. The Long Passage from Bellamy's to the Smoking Room communicated directly with the House of Commons offices. This whole range was consumed, and nothing was left but the walls by 11 p.m.: 'the flames raged very furiously at the corner constituting the Members' waiting room, and above *Bellamy's*, and when the roof at last fell in, the fire was so voluminous as apparently to threaten the contiguous building which forms the Committee rooms, and entrances for the Judges and Counsel to the Courts'.[5] Mr Bellamy, the caterer and resident housekeeper to the Commons, was not at home.[6]

Bellamy's was a famous Parliamentary institution. John Bellamy, its founder, was a wine merchant. He had been approached in 1773 by a group of MPs to establish a dining room in the Palace. This he did in two small rooms by the Court of Requests (later used as the House of Lords), serving delicious food and drink, which—as a canny businessman—was charged accordingly. Pitt the Younger's apocryphal last words were, 'I think I could eat one of *Bellamy's* pork pies', although some sources claim he asked for a veal pie. (Maybe the Prime Minister wanted a veal and ham pie: we will never know.) Bellamy realized that the unpalatable alternative for Members was to go to nearby chop-houses and taverns, which did not provide the collegiate atmosphere or top quality they wanted. Prices were therefore reasonable for London: a plate of bread, cheese, cold meat, and beer was 2*s*. 6*d*.; a full dinner of steak, veal pie, and chops with tart, salad, pickles, beer, and toasted cheese, 5*s*. 6*d*.; though claret—always a favoured Parliamentary tipple—cost an outrageous 10*s*. a bottle.[7]

Bellamy's son, the owner in 1834, had been appointed as Deputy Housekeeper to the Commons in 1811, jointly with his wife Susan. As in the Lords, the office of Housekeeper was a sinecure, so it was the Deputy who did the work in reality. However, the role of the Commons' Deputy Housekeeper was very wide-ranging and, as a result, extremely lucrative. The job did include a basic salary with a top-up for servicing committees and cleaning the Speaker's office; payment for managing stocks of stationery and sending off the post; a cut of the fees paid by the public to the Serjeant at Arms; an allowance for serv-ants; free gifts of old stationery; a complimentary set of the Commons main publications; and gratuities from Parliamentary agents and elec-tion committees. The net value of this, after payments had been made to his housekeeping staff, was over £429 a year.[8] On top of that, the family occupied a suite of rooms in the Palace and there were whatever profits the Bellamys were making from turnover in the refreshment rooms, which must have been very considerable. Some reforms to the system proposed in 1833 (as part of the investigation into fees and sine-cures in the Commons) had fizzled out so, altogether, the Bellamys had a lucrative little business on their hands at the time of the fire.

In 1834, Bellamy's was situated on the second floor of the Commons opposite the Abbey. The suite of rooms comprised a combined kitchen and refreshment room (what today we would call a snack bar), a Mem-bers' dining room, and a tearoom, with an adjacent storeroom. Com-mittee room no. 5 was next door. In the attic above were the family's private apartments—nine rooms with dormer windows in the stone

building, then extending down a passage to a large sitting room with a
further two bedrooms over, abutting the roof of St Stephen's.[9] Dickens
described how Bellamy's was the place of resort for all Parliamentarians, being

> common to both Houses of Parliament, where Ministerialists and Oppositionists, Whigs and Tories, Radicals, Peers, and Destructives, strangers from the gallery, and the more favoured strangers from below the bar, are alike at liberty to resort; where divers honourable members prove their perfect independence by remaining during the whole of a heavy debate, solacing themselves with the creature comforts; and whence they are summoned by whippers-in, when the House is on the point of dividing.[10]

On the night of the fire, Mr and Mrs Bellamy's sons managed to seize
the family's most precious books and papers—presumably the account
ledgers and business files that their livelihood depended on—before the
fire consumed the refreshment rooms, but 'to the entire abandonment
of their own property' in the apartment above. All of the Bellamys'
household goods were burnt—furniture, linen, 'wearing apparel' (both
theirs and their sons), plate, books, watches, trinkets, china, glass, pictures, and prints, costing nearly £1,843—and all the staff uniforms
were lost.[11]

Two of those staff were known to Dickens. Nicholas, the butler, occupied a pantry with a glazed hatch opening into the refreshment
room, from where he would loftily patronize junior MPs:

> He has held the same place, dressed exactly in the same manner, and said precisely the same things, ever since the oldest of its present visitors can remember. An excellent servant Nicholas is—an unrivalled compounder of salad-dressing—an admirable preparer of soda-water and lemon—a special mixer of cold grog and punch—and, above all, an unequalled judge of cheese. We needn't tell you all this, however, for if you have an atom of observation, one glance at his sleek, knowing-looking head and face—his prim white neckerchief, with the wooden tie into which it has been regularly folded for twenty years past, merging by imperceptible degrees into a small-plaited shirt-frill—and his

comfortable-looking form encased in a well-brushed suit of black—would give you a better idea of his real character than a column of our poor description could convey.

Jane, a waitress—'the Hebe of *Bellamy's*'—displayed a thoroughgoing contempt for most of the punters, though she loved flirting with handsome MPs, which 'you cannot fail to observe, if you mark the glee with which she listens to something the young Member near her mutters somewhat unintelligibly in her ear (for his speech is rather thick from some cause or other), and how playfully she digs the handle of a fork into the arm with which he detains her, by way of reply'. Jane's sarcastic remarks to guests were tossed out 'with a degree of liberality and total absence of reserve or constraint, which occasionally excites no small amazement in the minds of Strangers'. Her jokes and friskiness were impassively indulged by Nicholas: 'not the least amusing part of his character'. After the fire, Dickens rather poignantly reminisced:

> A queer old fellow is Nicholas, and as completely a part of the building as the House itself. We wonder he ever left the old place, and fully expected to see in the papers, the morning after the fire, a pathetic account of an old gentleman in black, of decent appearance, who was seen at one of the upper windows when the flames were at their height, and declared his resolute intention of falling with the floor. He must have been got out by force.[12]

Henry Taylor and his friend Edward Villiers had dined with John Rickman at the Athenaeum in Pall Mall in the late afternoon. As a good friend of the Clerk Assistant, Taylor had hurried with Villiers the short distance to Westminster on hearing of the fire soon after its outbreak, to see if assistance was needed, knowing that Rickman's house stood in the path of immediate danger.[13] As senior clerk in charge of the West Indian portfolio at the Colonial Office, Henry Taylor was busily playing a significant role in the affairs of the Caribbean colonies. Slavery had just been abolished, and there was much to do. He was very handsome, rather introverted, a chronic asthmatic—and was also meeting with some critical acclaim for his plays and poetry, the latter in the style of Byron. The Rickmans' proximity to both Westminster Hall and the Speaker's, with the wind blowing fully in their direction, had left other friends in no doubt that their 'utter destruction' was assured. One later told them that the inferno was so bright he could clearly see the points of the pinnacles of the west towers of the Abbey from his house in Acton, some eight miles away.[14] On knocking at the door and entering the house in New Palace Yard, 'such a scene of confusion never was seen', declared Villiers.[15]

The two men immediately took charge of the heavy operations, under the command of the capable Frances Rickman. Clothes were thrown into trunks haphazardly—light summer muslins squashed under heavy winter woollens and velvets, dresses stuffed into carriage boxes, and where keys could not be found quickly to retrieve clothing, whole chests of drawers and their contents were taken outside. Sheets and blankets were stripped from the beds and pressed into service as the family's possessions were swept into them, tied up, and transported out of the front door. Everything except the curtains and carpets was removed: including their books, pictures, and most of the furniture. Susannah Rickman's initial agitation and fear had subsided when she began to be active and useful. She took responsibility for packing her husband's clothes and daughter Anne's dresses, several chests of linen, the glass shades of the oil lamps, and her son's curiosity collection. Henry Taylor, looking horrified at the quantities of old porcelain in the house that needed to be removed, asked her if she valued it and when she said yes, Mrs Rickman found that 'a great Basket instantly was at my feet, and I packed the China'. The piano was taken off its legs and put in the hallway close to the door for a speedy exit, a nearby policeman looking eager to carry it off in his own arms. There it joined the mattresses, trussed up, and ready to move at any moment. A table tipped over and badly bruised Frances's leg, but she kept on working and giving orders all night long in a valiant effort to save her family's home.[16] 'In the midst of our agony', her mother later recalled, they were 'preparing for our own destruction'.[17]

At about half past nine an immense column of flame burst up through the roof and windows of the House of Lords, stunning the crowd:

> the whole of the upper part of the building was veiled from the eye by this outbreak of the conflagration; bright blue coruscations, as of electric fire, played in the volume of flames, and so struck were the bye-standers with the grandeur of the sight at this moment, that they involuntarily (and from no bad feeling) clapped their hands.

The blue electric fire perhaps proceeded, thought the writer of a letter to the *Gentleman's Magazine*, 'from the colouring matter in the materials of the tapestry of the House of Lords, which represented the destruction of the Armada, and the heroes of that event'.[18] It was a romantic but fanciful notion by that hour. The 'passive walls' of the House of Lords Chamber, the *Ipswich Journal* informed its eager readers, 'displayed an incentive to national vanity, by showing in needlework, never again to be imitated, the utility of that scheme of conquest...this elaborate work, the result of countless hours of solitary toil, is gone with the rest, and to history alone is now left the cold task of recounting the deeds of those heroes who there shone upon mimic canvas'.[19]

Contrary to popular belief today about the 1834 fire, the whole crowd did not stand around clapping and cheering in delight as the Houses of Parliament perished. The situation was much more complex than that. In reality it provoked a thoroughly mixed response: 'We saw the look of terror, of wild enjoyment, of wonder, and of admiration, of the different spectators,' said the *Metropolitan Magazine*.[20] Echoing the words of other newspapers of the time, George Manby described it as 'the most awful and dreadful conflagration I ever beheld'.[21] Today, we have lost the full force of the meaning which those words had in 1834. They mean that the spectacle was almost indescribable in terms of the amazement and fear it inflicted on the crowds. Contemporaries said the fire was (amongst other things) 'fearfully imposing' and 'the acme of sublime terror'.[22] Sometimes the effect was chilling, even unearthly. In the darkness, viewed from the white-stuccoed *palazzi* overlooking the trees across St James's Park, 'the many and monster-tongued flames...seemed to be moving and twisting, seeking fresh objects to involve in the ravening jaws of destruction'. The moon turned blood-red as fire-charged smoke ran across its face. The inhuman sounds of the wind and the flames were 'deeply awful, and seemed as distinct from the vulgar clatter and clamour of the mob, as do the solemn notes of the cathedral organ from the squeaking of the itinerant hurdy-gurdy'.[23]

So, the atmosphere was generally one of stunned astonishment, rather than festivity. The newspapers certainly did retell jokes made on the night, but to what extent the puns were real, or apocryphal to help circulation, is not clear. John Rickman marvelled at how the crowds were greatly interested in the fate of Westminster Hall in particular, and 'exulted loudly' when 'the engines were seen to prevail' and how the police and military maintained order without difficulty, 'no outrage attempted'.[24] In his private diary, Lord Broughton recorded how

> The crowd behaved very well; only one man was taken up for huzzaing when the flames increased. I heard nothing of the exclamations recorded by the *Standard* newspaper, but I believe that one weaver did say, 'This comes of making the poor girls pay for their children'—alluding to the new Poor Law. A few persons attributed the fire to design, but, on the whole, it was impossible for any large assemblage of people to behave better.[25]

Many journalists were also at pains to point out how well the crowds behaved. *The Times* was eager to reassure its readers that it did not 'observe in more than one or two instances any expressions of levity, and not one of exultation...the general feeling seemed to be that of sorrow, manifested either by thoughtful silence or by occasional exclamations of regret'.[26] Meanwhile, the *Manchester Times & Gazette* declared that the fire provoked 'feelings in which sorrow, astonishment and doubt will be singularly mingled'.[27] There appeared, on the part of the people, wrote the *Bristol Mercury* firmly, 'no disposition whatever to riot'.[28]

There was, in fact, a seeming obsession found in contemporary accounts—in newspapers, and in the comments of the political and literary classes—with the good behaviour of the crowd at the fire. The effectiveness of the arrangements made by the soldiers and police were particularly noted. By nine o'clock three whole regiments of Guards were on the spot under the command of Sir George Hill, Colonel Woodford, Lord Butler, and Captain Davis. Among the dignitaries joining Melbourne, Brougham the Lord Chancellor, and Duncannon, were three of the King's sons: the Earl of Munster, and Lords Adolphus and Frederick FitzClarence; and these 'distinguished characters were actively engaged in managing the military and the police', the latter being superintended by Mr May, of the A division.[29]

The reasons for this cautiousness are not difficult to fathom. Even forty years on from the Terror, there was still a nervousness in England about any massed gathering, especially those political in nature. Fear of the mob, and what it might do, lurked only just below the surface of the political classes at the time. The European revolutions in Europe in 1830 were very fresh in the mind, while at home the Swing riots of the same year had considerably shaken the government. The agitation surrounding Catholic Emancipation and pressure for the Reform bill had also been deeply disturbing for those in power. Lord Grey had written to a friend in March 1830, 'ministers are deceiving themselves very fatally as to the real situation in the country and the spirit rising within it. Another year like the last and who can answer for the consequences?' The following month, he wrote even more explicitly that 'this is all too like what took place in France before the French Revolution'.[30]

The crowd may also have had some misgivings about being there in large numbers, especially as the crush increased. Memories of the Peterloo Massacre of 1819, one of the most notorious events in the years leading up to the Great Reform Act, still lurked within memories of many. On that occasion, there had been eight fatalities, and 654 casualties who sustained at least 802 recorded injuries between them, among the peaceful masses.[31] But it was the more recent riots attending the passage of the Great Reform Act which would have been at the forefront of many people's minds. Once the news got out that the bill had been defeated during its first attempted passage through the Lords on 8 October 1831 there was uproar in the country. In Birmingham the bells tolled all night, and 100,000 people attended a protest meeting led by Thomas Attwood. Anti-reforming peers and bishops who had spoken against the bill were threatened in the street, some were even attacked. The *Morning Chronicle* and the *Sun* papers were both printed with black borders the following day, and on 12 October 1831, 70,000 people marched through the West End to present an address in support of the Whig ministry and the bill to the King. Political protest unions along the lines of Attwood's Birmingham Union sprang up. There were serious riots in Derby and Nottingham, but the worst one was in Bristol, where Charles Wetherell, an anti-reform MP, bullishly arrived on 29 October to hear a Gaol Delivery in his role as Recorder of Bristol. The Mansion House where he was lodged was besieged, and Wetherell had to escape over the roof for fear of a lynching. A full-scale riot ensued over the next three days, with houses burnt and pillaged, culminating in a troop charge which resulted in at least 400 casualties and an unknown number of fatalities. The military commander who lost control of the situation later committed suicide.[32]

Nerves were therefore jangling on all sides in the early 1830s. Fear of the authorities, and what they might unexpectedly do with swords and bullets to large masses of over-excited people, accounts for the panic among onlookers which had followed the collapse of St Stephen's an hour earlier when some in the crowd thought they were being fired on. Added to this tension, a potent cocktail of adrenaline, fear, excitement, noise, chaos, spectacle, the sense of night-turned-to-day and (by mid-evening) the certainty of the old Palace's destruction nevertheless did lead to jocularity among some in the crowd; even a sense of hilarity. Black humour made sensational and titillating copy for the papers, which recited each incident with evident glee. The levity of the comments was 'disgusting in the extreme' according to the *Bristol Mercury*, including those emanating from persons of apparent respectability, such as:

> 'Tis a pity the Bishops are not in it'—'Mr Hume's motion for a new house of commons will now be carried without division'. A great number of noblemen and gentlemen went in their carriages to see the spectacle, which all seemed to regret, except some of the lowest class of beings, who appeared quite delighted at the prospect of what they called 'benefit to trade'.[33]

Some, it was said, saw 'nothing in the sublimity before them but a huge bonfire', enjoyed the discomfort of the upper classes, and 'only wished that the fire might extend more and more, in order to heighten their enjoyment'. However, these comments came, the *Metropolitan Magazine* reassured its readers, only 'from the most depraved denizens of the most depraved quarter of a vast metropolis'.[34] The immediate thoughts of the crowd seem not to have related to the Great Reform Act, as emerged later, but to the most recent legislation of that year. 'There's a bonfire for the Poor Law Bill!' one person was heard to exclaim, while another joked how the House had risen, 'amidst great uproar and cries of Oh! Oh!' A man with a placard displaying a sensational picture of the horrors of the Poor Law stationed himself by the Canning statue and filled his pockets by selling abstracts of the bill.[35] An elderly well-to-do lady waxed eloquent on her lack of regret at the fire, given that 'the abominable Poor Law Bill is burnt'.[36] A newcomer contemplating the fire for a few minutes exclaimed, 'Well, I'm blessed if I ever saw such a flare-up as this before', and his neighbour allegedly replied, 'Nor I! I never thought the two Houses would go so near to set the Thames on fire.'[37] The historian and critic Thomas Carlyle was in the crowd, and heard similar comments, suggesting not all were apocryphal:

> I saw the fire of the two Parliament Houses; and what was curious enough, Matthew Allan (of York, you may remember) found me out in

the crowd there, whom I had not seen for years. The crowd was quiet, rather pleased than otherwise; whew'd and whistled when the breeze came as if to encourage it: 'there's a flare-up (what we call *shine*) for the House o' Lords'—'a judgement on the Poor-Law bill!'—'there go the Hacts (*acts*)!' Such exclamations seem to be the prevailing ones. A man sorry I did not anywhere see.[38]

Curiously, in 1834, Carlyle had just begun to write *The French Revolution: A History* (1837)—so perhaps thoughts of the power of the mob passed through his mind that night as he too stood in the peaceful crowd. (Ironically, the first draft of this manuscript was itself accidentally burnt six months later by a servant of the philosopher J. S. Mill, to whom he had lent the draft, and who had taken it for waste paper.)[39] Some incidents were sadder. A respectable, but confused, elderly man called James Brown was arrested for 'conducting himself in the most violent manner' and cheering in delight, 'This is what we wanted—this ought to have happened years ago!' and other expressions of a similar nature. He was restrained (only with some difficulty) by the police from running into the flames, and was taken into safe custody until his friends could be found.[40] A drunken coalheaver (or 'black diamond merchant' as he was described) tried to pass through the ranks of soldiers between Abingdon Street and Old Palace Yard and was stopped. An altercation took place between him and a soldier, as he asked the 'fine lobster' why he couldn't 'go and see my own property a-burning?' When challenged on this he replied bitterly that he regarded it as his since no doubt 'they'll lay a tax upon me for to help build it up again'.[41]

The *Morning Herald* published a leader which may have led later generations to assume that the entire crowd was rejoicing at the fire of 1834. 'The mob,' it wrote (use of the word 'mob' here is of course interesting), 'upon witnessing the progress of the flames, raised a savage shout of exultation; but, unreflecting people! what would they have been had not the very identical House of Commons over whose fall they triumphed, existed, but mere serfs and slaves?'[42] William Cobbett then obtained a copy of the paper on 20 October while staying in Limerick, and the response of the well-known radical to it immortalized the *Herald*'s copy. Cobbett described in his own paper, the *Weekly Register*, how he had read

> an account of the BURNING of the Parliament House! As to the CAUSE whether by fire and brimstone from heaven, or by the less sublime agency of 'SWING' my friend, the Herald, does not tell me; though this is a very *interesting* portion of the event...my insipid friend [the Herald] says, 'that the MOB (meaning the people of London) when they saw the progress of the flames, raised a SAVAGE shout of EXULTATION' Did they indeed! The Herald exclaims 'O UNRE-FLECTING people!' Now perhaps the MOB exulted because the MOB was really a *reflecting* 'mob'. When even a dog, or a horse, receives any treatment that it does not like, it always shuns *the place* where it got such treatment: shoot at and wound a hare from out of a hedge-row, she will always shun that spot: cut a stick out of a coppice and beat a boy with it, and he will wish the coppice at the devil: send a man, for writing notorious truth, out of the King's Bench to a jail, and there put him half to death, and he will not cry his eyes out if he happen to hear that court is no more. In short, there is always a connexion in our minds between the sufferings that we undergo and *the place* in which they are inflicted or in which they originate. And this 'unreflecting mob' might in this case have reflected that in the building which they then saw in flames, the following, amongst many other things took place...

Cobbett then goes on to cite a list of laws passed by Parliament since the Reformation which he deems oppressive and cruel, including the suppression of Catholicism and introduction of the penal laws, the Riot Act and the Septennial Act, wars against America, treason, trespass and felony laws, and the rights to tithes applied by the Anglican Church. 'But I must break off', he writes, exhausted at the end of his rant. 'The post is going. I will finish the list next week.'[43]

Cobbett also reprinted a passage from the *Standard* which pronounced that

> The sense that the property of respectable persons was in course of ruin and that the lives of many brave and honest men were in jeopardy, alone controlled a universal disposition to merriment. Hundreds

confessed the feeling (*of course jocularly*), avowed by the old Earl of Kildare, when he apologized for burning Cashel Cathedral, by solemnly protesting his belief that the Archbishop was in it. Some wags said they hoped the fire would reach the Poor-laws Amendment bill. Others regretted that the bu[l]k and lumber had not been burned years ago. There was no mischief whatever in these pleasantries...but if ever the suffrages of a whole city were unanimous upon one point they were last night given to this, that there has been nothing in the existing generation of parliamentary men to command veneration or even respect. The burning of an oil store in Thomas-street, a few nights ago, excited solemn and sympathetic feelings exactly the opposite to the predominant sentiment in Palace Yard, last night...Instead of regretting the event as a national calamity, many appeared to consider it as a well-merited visitation, and actually openly expressed their regret that the Lords and Commons were not sitting at the time...This spirit, however, we lament to say, and we speak from personal knowledge, was not confined to the lowest and most ignorant order. Many individuals well-dressed, openly professed to feel but little regret while witnessing the progress of the flames. They seemed to think that a visitation so awful would induce the legislature to adopt some different measures to those that have latterly occupied their almost exclusive attention; and any event that would produce that effect, they were inclined to consider as a special interposition of Providence.[44]

In quoting the *Standard* however, Cobbett had deliberately chosen one of the more inflammatory rags to report, to boost the account of his 'insipid friend'. For example, here is another passage from the *Standard*, which casts doubt on the paper's overall reliability:

The pretty general impression yet is, that this dreadful event was the work of an incendiary. There is no doubt but that the fire commenced in the passage called the 'Bishops Lobby' where for some days past several plumbers have been employed opening and clearing the flues which communicate with the House of Lords. It is not however, yet discovered how they could possibly have even carelessly occasioned the mischief.[45]

One thing is certain. For all the newspapers and magazines—Tory or Whig, pro-establishment or radical, lofty or scurrilous—it was the story of a lifetime.

That wry sense of just deserts having been meted out on politicians, indeed a sense of divine intervention having been involved,
was also reflected in comments such as those of the piously evangelical
George Crewe, squire of Calke Abbey in Derbyshire, who wrote in
his diary on 19 October on hearing the news: 'at present we know
nothing, perhaps never shall know any theory of the real origin of the
conflagration...It matters not, perhaps, by what agency this awful
visitation was accomplished. A visitation we must consider it, and in
these times of trial and of trouble, one which is ominous of the existence of threatening danger, the heavy hand of which is gathering
over our heads.'[46]

For some, it must indeed have seemed that the Apocalypse was near.
St Stephen's continued to burn away to its medieval core. At least one
report of the effects of the fire on the Commons Chamber considered
the sacrificial, purifying aspect of the blaze, almost as if the authentic
House of Commons were rising, phoenix-like, from the ashes of its
own funeral pyre:

> The old mouldings and carved stonework of which, by the burning out
> of the wooden box, or lining, which formed the House of Commons,
> are to be seen to great advantage...Its aspect within is very curious and
> interesting. Of the House of Commons, which had been obtruded into
> this building with a barbarism of taste that would have disgraced the
> meridian of Constantinople, nothing now remains but rubbish and rot-
> tenness and smoking embers. The whole of the intrusive arrangements
> which converted the finest chapel in the kingdom into the worst
> imaginable Chamber of Legislation have withered away like a burnt
> scroll, and revealed the original walls and proportions of the building,
> with much of the original mouldings and tracery—the carvings and
> paintings with which they were decorated. It is really wonderful to see
> the sharpness and beautiful finish of the mouldings, the crokets, the
> embossed ornaments, and cunning workmanship in stone, notwith-
> standing the violence which the Chapel had suffered from ancient de-
> stroyers and modern improvers, besides having just come out of the
> fiery furnace of so tremendous a conflagration. The fire is not so cruel
> then as some people have supposed.[47]

The antiquarian longing expressed in this passage for a lost architectural past was a foretaste of the later demand, laid down in the competition for a new Palace, for a replacement building only in the Gothic or Elizabethan style. By its very design, a new home for the House of Commons and House of Lords would somehow provide a fresh start for Parliament: cleansed of its murky and labyrinthine ways; untainted, honest, and free of the corruption of later ages which had weighed it down. From Lambeth Palace on the other side of the river,

> more of the detail of destruction was visible...the gable end of [the House of Commons], the last relic of the Chapel of St Stephen, with its beautiful gothic window stood up between us and the glowing furnace of fire in strong and beautiful outline. Whilst at intervals, as the smoke varied in its motion, other portions of gothic architecture, oriel windows, turrets, and towers broke on the view; and all within side them was feeding the devouring element. To the right of these a cluster of ancient elms intercepted the view, as if Nature would screen, in her sorrow, the destruction of Art; and to the left, the long and lofty roof of Westminster Hall stood forth like a line drawn between ruin and protection. As the night grew and the moon attained her meridian, and as the flames made greater progress, singular and sometimes beautiful changes took place, in the appearance of all around.

The huge column of flame over the Palace was

> gradually losing its power, and following the direction of the gale, it bent forwards towards the east, and extending far across the river, covered the opposite shore in a wreath of murky smoke, lit up here and there by flakes of fire, that sometimes fell so thickly as to excite apprehension for many of the slightly constructed and wood-built houses along shore. On each side of the river the reflection on the windows of hundreds of the more prominent of the houses was so strong, that, shielding all else from the sight, it seemed as if some joyous event were celebrating by a general illumination. Now and then, as the immense flame veered a point or two more northward, the Abbey appeared, darkling and sadly in the dim distance.[48]

In the days that followed, a quiet acceptance and pragmatism seems to have prevailed. 'I thought the old House so ill adapted to its purpose', mused the MP (and pioneer photographer) W. H. Fox-Talbot on hearing the news while on holiday abroad, 'that I cannot help approving the notion of our having a new one, more especially now that we have no choice on the subject.'[49] The question of a new House of Commons, declared the *Bristol Mercury*, 'has been settled in a manner unexpected by any and, we should hope, regretted by every one'.[50] But it was Lord Althorp, the Chancellor of the Exchequer, who had the last word in pragmatism. He had cried out in desperation, for which he was later censured by some: 'D—n the House of Commons, let it blaze away; but save, oh save the Hall!'[51]

9

But save, oh save the Hall!

THE STRUGGLE TO do just that had been going on for several hours, but it began to reach crisis point from ten o'clock onwards:

The [north] front of Westminster Hall was the grandest feature, where all was painfully grand. It stood up in mourning blackness, as a barrier, beautiful, high and vast, that kept back the flames from rushing upon the multitude. Through its fine gothic window, in all its maze of tracery, we saw the ignited rafters at the upper end, the [north] window itself appearing one blaze of light, with every compartment marked out by the densest darkness. The large door below the window was opened, but you saw nothing within, but what appeared to be an unfathomable mistiness of fire. It was as if one looked into a vast furnace, undistinguishable, and filled only with the most consuming heat. Upon the low roofs on the left [east] of the Hall, little black things were seen moving busily along, appearing so small and insignificant in that gigantic struggle between the dread angel of fire and the genius of antiquity, that we almost blushed to acknowledge them to be men.[1]

It may have been this seemingly inevitable destruction which was said to have provoked Melbourne, at a quarter to the hour, to direct the police and military to throw all the papers and boxes out of Soane's Law Courts into the street: 'and in a few minutes the windows were smashed in, and the streets covered with papers and rubbish of every description', though in fact these rescue attempts had begun much earlier on.[2] The Hall's interior had not been so splendidly illuminated since the coronation banquet of George IV in 1821. This time, though, it was a sinister floodlighting: one that indicated disaster was imminent. The strong light flickered across its gorgeous roof trusses, while twenty-six giant wooden angels, holding carved escutcheons in front of them as if to protect themselves from the intense heat, looked down impassively on the scene of confusion below. Even those spectators who knew little and cared less about the other Palace buildings were worried: 'the very mob seemed to care little for the destruction of the other buildings on which they vented their low and reckless jests, but the feeling of anxiety was almost universal for the preservation of the noble Hall'.[3] Nearly 750 years old at the time of the fire, it was the most visible part of the medieval Palace and certainly its most ancient. It had a special place in the national consciousness: 'for the beauty of its architecture and its close connection with some of the most important events of our country's history, [it] is equally admired and estimated by the antiquarian, the man of science and the citizen'.[4] Certainly the 'utmost anxiety prevailed' for its survival, particularly when onlookers realized that there was wooden scaffolding inside which might catch light at any time.[5] The fight to save the Hall was undertaken by many uncoordinated groups, without a single controlling intelligence. The accounts of events are disjointed, confusing, and contradictory. On one thing, however, they all agree, and that is that the effort was fiercely determined, dangerous, utterly exhausting, and took place mainly between eight o'clock and the early hours, as hundreds worked tirelessly to keep the Hall and its magnificent roof from being incinerated.

James Braidwood was not in a position (as superintendent of a private, rather than public, metropolitan fire service) to take overall charge. During the course of the evening, the Prime Minister, various members of the Cabinet, royalty, several bodies of guards with their commanders, numerous firemen, and many civilian volunteers appeared on the scene, so 'there was zealous interference now on all sides', wrote the *Examiner*, but 'great want of a commander in chief '.[6] On the other hand, another onlooker, Charles Fremantle (a naval officer), viewed the chaos and complete lack of leadership and wrote in disgust to his brother Tom, Tory MP for Buckingham, how many of the Ministers on the spot, including Melbourne, Palmerston, Althorp, Auckland, and Hobhouse, were 'all standing like stuck pigs without attempting to do anything to prevent the confusion & mismanagement, which was great'.[7] All the great and the good—Lord Hill and the Earl of Munster among them—were in a state of extreme agitation for the Hall, as was Mr Westmacott (the sculptor of the Canning statue outside), and a cry was heard when an assistant engine was called for, 'What is one, bring half a dozen!'[8] The heads of various fire insurance offices, plus Messrs Lott, Merryweather, and Bristow, the well-known fire engine makers, were also in attendance, rendering 'great assistance'.[9]

First, then, the LFEE firemen's experiences from the time of their arrival onwards. We are fortunate to have not only Braidwood's accounts of the fire, but also a detailed idea of his methods. These come from his published manuals on firefighting: *On the construction of fire-engines and apparatus, the training of firemen and the method of proceeding in cases of fire* (Edinburgh, 1830) and a posthumous collection of his writings entitled, *Fire Prevention and Fire Extinction* (London, 1866). The first treatise would no doubt have influenced proceedings in 1834; but the Westminster fire was so enormous that it also in turn must have shaped the further development of British firefighting theory and practice outlined in 1866. Much of what Braidwood has to say later certainly chimes with what is known of his experiences and actions during that night.

139

Braidwood's ideal fire engine was a design manufactured by W. J. Tilley, of 166 Blackfriars Road, London, based on the fire chief's theories developed in Edinburgh, and then honed to perfection in London. In 1830, he had felt that a small, twelve-man engine throwing 41 gallons a minute from twenty-four strokes of the levers was the ideal, particularly in terms of manoeuvrability; but later in life he came round to the view that a larger engine was better, though it still needed to be easily shifted as required. This was surely influenced in some part by the Westminster conflagration. An engine of 17½ hundredweight (890kg), when loaded with another four hundredweight of hoses and tools, was quite enough for two horses to pull for a distance of six miles; particularly when topped off with five firemen and a driver. This threw 88 gallons (400 litres) when worked off forty strokes of the levers a minute: three or four hours' work required a team of twenty-six men, maybe thirty if the hoses were far from the fire-plug and required more exertion.[10]

Fire engines were handsome beasts. They had cisterns (tanks) of oak or mahogany, with lighter Baltic fir uppers, inside which was a valve and suction mechanism of cast iron and shiny brass. The lever shafts running the length of each side were made of lightweight lancewood to aid the pumpers' action, and folded inwards when not in use or being transported. The wheels were sprung, enabling the engines to be pulled over rough roads or cobbles (which was especially important in Edinburgh). These were the kind of engines in use by the LFEE on the night of the Westminster fire. The equipment which accessorized each engine was extensive. It comprised seven 40-foot hoses, made of leather with copper rivets; a bundle of sheepskin strips for emergency hose repairs; four suction pipes (six to seven feet in length, to get the water into the engine); two copper 'branch' pipes (four feet long and one foot long, for exterior and interior work respectively); three jet pipes or nozzles; three wrenches; two lamps; two lengths of scaling ladder (each six and a half feet long which could be attached to others to create a ladder seven or eight times the length); one fire hook (for pulling down ceilings); 60 feet of patent

line (for hoisting hoses through windows from outside rather than dragging them up stairs); one mattock and one shovel for damming water
and uncovering drains; one hatchet or pole-axe; one saw (for cutting
rafters, beams, and other structural woodwork); one iron crowbar; one
portable cistern (a square canvas bag stretched over a folding iron frame,
which could fit over gushing street-plugs to form a well of water into
which two or even three engines could immerse their suction pipes); one
flat suction strainer (to prevent stones and debris being sucked into the
engine with the water from the plugs); one standcock and hook (this
could be pushed directly into a street-plug with a hose on the other end
to throw a jet of cooling water on ruins); one screw wrench; and one
large canvas sheet with ten or twelve rope handles round it (to catch
people jumping from windows). Finally, the ideal engine included a
handpump, with 10 feet of hose, a jet pipe, and nine canvas buckets.
This could throw 6 to 8 gallons to 30 or 40 feet: very useful for keeping
doors and windows cool.

Braidwood believed that fire engines should be placed in a straight
line between the supply of water and the premises on fire, but the
pumpers needed to be far enough away not to be scorched, or annoyed
by falling water. Placing engines too close to a blaze was pointless.[11] So,
on his orders, probably sometime between his arrival with the rest of
the brigade at half past seven, and eight o'clock, a street-plug was
drawn in New Palace Yard, outside the north end of the Hall. Two of
Braidwood's engines were manoeuvred into position, ready to take up
the water from the pipes. An onlooker describes one of the LFEE engines there, with its hose stretched into the Hall through the north door.
But at 239 feet (73m) long the Hall was so enormous that the hose
proved too short to stretch to the south window at the far end. An extension hose was therefore dragged into the Hall by two firemen, Mr
West and Mr Williams, and attached to a second fire pump stationed in
the centre of the Hall itself under the command of Mr Tree and Mr
Solomons. The two engines and two hoses therefore formed a single
super-unit for pumping the water, and there seem to have been two of

these units (four engines, four hoses) in use in the early stage of the fight. The rigid 'branches' of each of the hoses were then carried up ladders and through a window onto the leads, where the firemen began to aim them at the flames.[12] The 'judicious measure' of stationing engines in the interior, 'which could pour a stream of water on any part of the roof immediately threatened', was wholly sound, thought one onlooker.[13] But another observer noted the lack of organization: 'the ladders also having been quickly taken, under the direction, *still*, of *casual* advisers and placed against the great window, the firemen, ascending to its base, found ample room to play upon the blazing element which thence confronted them'.[14]

The scaffolding—erected by Robert Smirke as part of his restoration work for John Phipps at the Office of Works—instead of becoming a means of destruction, now unexpectedly worked to the rescuers' advantage. An atmospheric watercolour in the Palace of Westminster Art Collection shows the frantic activity in the Hall at this time, crowded with people, some spectating, others gathering round the engines, and still more shinning up the scaffolding on ladders. Many red-coated soldiers are heaving the engine levers. One man downs a large mug of beer in the foreground. Once the rescuers were at the top of the scaffolding, they were able to hoist up hoses, smash windows, and get out onto the roof. From there, they could pull the hoses through or start removing tiles to create a firebreak between the Hall's roof and the burning Commons buildings.[15] For an hour or so, these two super-units were sufficient. But once the conflagration had extended to the House of Commons on the right, and the buildings in the Speaker's Yard on the left, where two engines belonging to the LFEE and one to the Exchequer Court were in full operation, it was found necessary to bring another engine into the Hall immediately. The LFEE super-units were therefore further augmented by an engine from the County Fire Office, whose hose was snaked through a passage in the Hall's south-west corner, and also set to work in the same place.[16]

Braidwood was emphatic in his published theories that it was a complete waste of time and effort to throw water onto a surface of slate, plaster, or 'a dead wall'. The firemen needed to follow the blaze to its seat by means of the copper branch pipe at the hose's end—taking it upstairs, and down, in at windows and out through the roof, if necessary. 'The old plan of standing with the branch pipe in the street and throwing the water into the windows is a very random way of going to work,' he wrote, 'and for my own part, although I have seen it repeatedly tried, I never saw it attended to with success.'[17] The man who holds the branch pipe ('considered the post of honour') should 'get so near the fire, inside the house, *that the water from the branch may strike the burning materials*'. That water which does not strike the burning materials falls within the property; 'and by soaking those parts on which it falls, prevents their burning so rapidly when the flames approach them'.[18] And indeed, outside the Palace the old-fashioned, ineffective method of fighting the fire by the parish firemen was still being used, causing one journalist to complain that 'the whole should have been marshalled in a line. Little was done to direct the jets of water to the requisite openings, by persons raised on ladders to the necessary height, and consequently the greatest part of the water intended for the upper stories of the building did not reach the object, and was wasted.'[19]

The 'Jimmy Bradys' were immensely brave. Outside the great south window of the Hall was a small lead flat—a platform just a few feet square—over the passage that led from the Hall to the House of Commons on the principal floor. Balanced on this, four LFEE firefighters—Tree, Solomons, West, and Williams—played their hoses on any flames that came near, following Braidwood's techniques. At times they were surrounded on three sides, both around them and below, sometimes as close as two feet away. They and two or three assistants moved the hoses around as necessary to allow the water to be aimed where it was needed most.[20] For the first hour of their use, this seemed to hold things at bay and partly allayed the fear entertained that the fire would communicate to the great hammer beams.[21]

143

But then things took a turn for the worse, and the flames got so close to the roof that sparks were falling through the south-east corner of it into the body of the Hall. This can be seen in another thrilling watercolour by G. B. Campion, showing the exterior of the Hall almost floating on a river of steam and water, as orange flames attack one corner of the roof surrounded by men desperately trying to create a firebreak as fast as they can.[22] And the firemen continued to work, one spectator described,

> with all the energy of antiquarians who were sensible of the immi-
> nent danger that hung over a venerated pile, whose loss could never
> be restored; and their extreme tractability in directing the pipes of
> their engines from place to place, as an anxious bystander (who had
> better opportunities of making observations than themselves) wished
> them, was deserving of the most unqualified praise.[23]

Far below the firemen, on the flooded stone slabs of the Hall's paving, the engines provided a steady stream of water for hours on end, powered by multiple shifts of volunteer assistants. Braidwood's normal practice was to ensure that the fire-engine levers were pumped by bystanders paid one shilling for the first hour (sixpence for each succeeding hour), besides receiving liquid refreshment. Upwards of 600 assistants had been thus employed, wrote Braidwood in 1861.[24] He was perhaps thinking of the numbers that had been required in 1834: that evocative watercolour of the Hall interior suggests so, and he certainly later acknowledged the aid of 'numerous assistants' that night.[25] High and low together manned the pumps. Even the Attorney-General rolled up his sleeves and joined in, injuring himself: 'I received a wound on my knee—not seriously hurt—in working an engine to save Westminster Hall, which was twice in flames . . . between two and three . . . I am greatly delighted that the Hall was saved', he wrote breathlessly later that morning. The event was not an unqualified success for him, though: 'in addition to my wound, I had my pocket picked of a purse with four or five pounds and a pocket-handkerchief'.[26] Firemen (despite being of 'strong and commanding frame') were in danger of catching cold from being soaked through so Braidwood made it a rule to give each a dram of spirits two or three hours after a fire's commencement. Where a fire required that engines had to be left on the spot to continue prolonged work, the crew was sent home to change their clothes.[27] Mr Barber Beaumonth of the County Fire Office also helped to rally the exhausted firemen fighting in the Hall on the night, supplying them from time to time with necessary refreshments.[28]

As eleven approached the crisis reached its height. 'The flames were spreading so rapidly towards Westminster Hall that it appeared as if every minute it would fall a victim to the destructive element', wrote one.[29] The flames were attacking 'with great force' both flanks of the Hall—the committee rooms to the west, and on the other the private apartments and passages of the Speaker's House to the east.[30] From inside the Hall, the rescuers could see the destruction being wrought outside:

> The flames at each side showed ominously through the upper line of Gothic casements, flaring against the old oak timbers; and on the eastern side, next the Speaker's House, the fire seemed to glow through the lath and plaster with which they had been screened up; while, in one place, where there had been a private door, the wooden frame-work blazed round an orifice, which seemed like the mouth of one of the potteries. Here the Hall was, in truth, on fire. Before the great window, at the same time, there was a deep dull red, in the midst of which the ribs of the burnt building stood, but occasionally veiled by thick volumes of piles of brick, or a fall of burning particles. On the floor of the Hall, and amid piles of brick, newly-hewn stone, timbers, and all manner of obstructions, were the two engines worked by their respective companies' labourers.[31]

Having given up hope of helping at the front of the House of Lords, at some time after seven, Captain William Hook of the Royal Navy had made his way with some expedition round to the Speaker's Yard where he introduced himself to the Earl of Munster who was observing the scene.[32] Throughout the evening, the Earl was heard repeatedly to exclaim that he would make any sacrifice to save the Hall.[33] 'Who are you?', demanded Munster of Hook: a question that might equally well have been levelled at him. George Augustus FitzClarence, Earl of Munster (1794–1842), was the eldest son of William IV (formerly the Duke of Clarence) and his mistress the actress Dorothy Jordan; one of ten children produced by their twenty-year liaison. William IV had no legitimate children with his wife Queen Adelaide. Munster, a soldier, had had a somewhat mixed military career, distinguishing himself in the Peninsular War, and even becoming Wellington's adjutant-general, but he was headstrong and had then been involved in a mutiny against his commanding officer which led to a court martial. He was also obsessed with his illegitimacy and the status of his mother after she was cast off by William in 1811, with the prospect of his father's inheriting the throne in sight. There had then been a furious row when William refused to allow him to carry the crown in state at the coronation in 1830.[34]

Once Hook explained his position and skills, Munster replied, 'We are alright, and understand each other', but stated that he had no authority to act either. Hook explained to Munster that they needed another engine in Westminster Hall, and that spare hands should be employed under the police to remove the Parliamentary papers and records to the buildings opposite, a gentleman on the spot having obtained leave from the occupiers of the houses to lodge anything that could be saved. The King's son agreed to try to influence 'by my presence' anything that could be done for the best. So, fearing that the temporary scaffolding on the west side of the Hall might catch fire, Hook prevailed upon a master mason to order his men to start removing the part nearest the great south window, holding himself responsible for their action.[35] Different rescuers were now working—unwittingly—in opposition to one another.

Shortly afterwards, Munster, Lord Adolphus FitzClarence his brother, the Marquis of Worcester, a Captain Gordon, and a number of other noblemen and gentlemen (including Hook of the Royal Navy and Captain Thornton of Palace Yard) could be found conferring in the Hall, very anxious 'for the preservation of the ancient and splendid edifice'. A barrister in the group suggested they should try to see what progress the fire was making in the Commons corridor on the other side of the south wall. He was particularly worried that it might have reached the floors above, since the strong light through the great south window indicated that something ominous was afoot. Together, this band of intrepid gentlemen tried to open the locked door under the window, banging on it for a long time in an attempt to attract attention and gain admittance. Searches for any messenger or House servant with keys to the door also proved fruitless. Finally it was decided (desperate times calling for desperate measures) to break it open 'by main strength', using one of the scaffolding poles as a battering ram. As the group burst through into the passage on the other side, they found a large number of Acts of Parliament, reports of committees, and other papers, which were rapidly removed by all hands to the northern end of the Hall for safety. William Hook mounted by the nearest stairs towards the roof, and found that the fire was gaining rapidly towards the Law Courts. He ran down again to the parish and local fire engines, most of which were still playing uselessly over the frontage of the House of Lords, 'long past the prospect of recovery'. Despite begging the pump team leaders, he could not find anyone to go with him to see what he had found, until he was referred to one of the fire engineers (called Turner or Edwards, he thought), who went back with him, taking along a fireman. The engineer saw the danger, and Hook was ordered to stay put and await the engine's arrival so that he could direct where to position it, while the engineer gave the necessary orders on the ground.[36]

The Times corroborates Hook's exploits:

The fire had by this time made most fearful progress, the flames rising many yards above the burning buildings and myriads of sparks flying into the air: it was proposed to have the scaffolding, which had been erected for the repairs of the inside of the Hall, removed for fear of the fire communicating to it from the outside. This was opposed by

147

several gentlemen on the ground of it being useful to the firemen, in directing their operations and it was ordered to remain and by means of it the hose from two more engines were carried up on each side of the Hall to the outside of the roof, to which, there is no doubt it is to be attributed the preservation of the building. A door at the south-western end of the Hall was then forced open and Earl Munster &c went through the passages into the Commons lobby. Here the fire was fast descending from the upper part, to prevent which the breach from the British Engine was brought up the staircase, but notwithstanding the greatest exertions of the men, who under the direction of White the engineer were devoted to effect that object, it was found impossible to stop it, and they were compelled to retreat.[37]

This having been done, the indefatigable Hook then rallied together about twenty guardsmen and commenced unroofing the narrowest part of the Hall roof which adjoined the burning Commons buildings, 'in order to cut off the communication should the fire gain either before or after the engine'. Possibly hoping for some preferment, or perhaps simply wishing to provide information to the Home Secretary, Hook later wrote to Duncannon about their desperate efforts, which

continued until enough of the roof and the ceiling below was displaced to render our object nearly secure but immediately the engine had sufficiently subdued the fire, assisted by water past along in buckets from the cocks on the upper premises, the unroofing & breaking down farther of the ceiling was discontinued; and I do myself the honor of representing this, my Lord, as I hold myself entirely responsible for it, born out as I was by every gentleman assisting myself and the soldiers with a labouring man and that of one of the firemen who brought us the tools for the purpose and the presence shortly after of Lord Munster who witnessed the zeal of every one employed. In an hour or two the fire was sufficiently reduced in this quarter to enable us to leave it to the firemen and to direct our attention elsewhere.[38]

Thus—around eleven or midnight (it is hard to state an exact time due to the confusion)—Hook, accompanied by several others, descended into the body of the Hall, 'where the fire was also nearly subdued, except in one quarter'. Mounting a ladder, Hook battered in one of the windows on the east side; his group of soldiers cut through a roof, and then commenced taking off the tiles,

> until the engine in that quarter having slackened the fire, farther pro-
> ceeding was unnecessary but as the fire opposite on which the engine
> no 18 and the Speaker's engine should play, was gaining along a gal-
> lery at a rapid rate, there being no water to keep them going regularly
> I proposed to search the fire buckets from the Admiralty that the
> guardsmen might hand water from one to the other from the River,
> the tide being then on the rise.

A sergeant then told Hook that there were plenty of unused pails lying around the Speakers' Garden, and the two of them went with others to collect them.[39] The men lowered the buckets down from the iron rails at the edge of the garden to the water's edge and had begun to pass the water along a human chain when

> from some slight interruption from the lower classes below and the un-
> necessary interference of a police man stationed in the garden (and
> whose conduct I have represented to Colonel Rowan) we were com-
> pelled to knock off; tho' I have no hesitation in stating that after the line
> of men were once established (from the good feeling of two guardsmen)
> we should have succeeded at least in keeping one Engine going from this
> source. After this I remained at the spot till half past two o' clock when
> the fire was so far got under as to render any aid from volunteers
> unnecessary.[40]

Hook's exploits, of course, are the story of just one of hundreds of people in the Hall helping the rescue effort.

'On no account whatever', instructed Braidwood when training others to fight infernos, 'should directions be given to the firemen by any other individual while the superintendent of the brigade is present.' In that person's absence, it should be made clear who is in control.[41] This clearly did not occur in 1834. 'The firemen shouted their directions from above, and numerous busy meddling people, whose rank embarrassed, but whose wisdom afforded but little guide, [shouted] from below.'[42] All of this was in great contrast to activity elsewhere in the Hall, where an 'inactive but attentive' observer of the fire from 8 p.m. to 3 a.m. noticed that away from where the flames were getting most attention Lord Hill's calm, experienced judgement was being steadily directed to saving the Hall on another front. Rowland Hill was Wellington's favourite and most successful general, a brilliant and humane soldier who had risen to fame in the Peninsular War and who had attained the alarming distinction of having his horse shot from under him at Waterloo. He was said to give the second-best dinners in the Army, and was well known for taking an impressive silver service canteen with him on campaign to satisfy his epicurean tastes.[43] In 1834 he was Commander-in-Chief of the army. Unseen by almost everyone, Hill directed a fatigue party of the Guards to unroof a portion of the Speaker's House, 'by which means, in comparative darkness, he effected a gap to which the flames eventually arrived, and there stopped'. Had this decisive measure not been taken, the flames would have continued their progress along the roof and into Westminster Hall. He was one of the 'mildest, most modest, and most faithful servants of this country', concluded the onlooker.[44]

Another group of men, far more humble, but apparently extremely effective, were some thirty masons working under the direction of the contractor Mr Johnstone (Patrick Furlong's erstwhile employer). They had initially been taken on for the restoration works, but quickly assembled round him on the night of the fire, and to their 'zealous exertions' Robert Smirke's brother directly attributed the survival of the Hall. It is not exactly clear what they did, but it was crucial, and possibly associated with the scaffolding and access to the roof.[45] Johnstone may have been the master mason instructed—wrongly—by Hook to remove the scaffolding, earlier.

A different volunteer's experience seems typical, as he recalled how,

> a little after 7 o'clock on Thursday evening I reached Westminster Hall, when I found that part of the House of Commons contiguous to it one entire furnace of flames. I assisted some firemen in breaking down and taking away the large wooden doors in the communication between the two places in order that the flames might be more completely played upon by the water. Engines were also steadily directed against the southeastern and southwestern corners of the roof of the Hall adjoining the House of Commons, and the preservation of the former was, I think, in a great measure attributable to the constant manner in which these engines were kept at work. Some noblemen were on the spot superintending—I think Lords Hill and Althorp.

He then went off to help with the record salvage operations in the committee rooms.[46]

While the Hall was a scene of frenzied activity, some other buildings were being left to perish beyond all hope of recovery. As a result, 'between ten and eleven two great masses of the frontage of the House of Lords fell in, but in consequence of the heaviness of its timbers and, probably, its numerous mural subdivisions, it still continued to burn most fiercely'.[47]

At ten o'clock the 'Blues' of the Royal Horse Guards arrived from the Regent's Park barracks, at a time when the conflagration was raging 'with unabating fury'. Nothing, it was said, could exceed the praiseworthy conduct of the firemen and military. By half past ten most of the furniture of the Law Courts, including the Exchequer, had been moved to the pavement nearly opposite the entrance of Poets' Corner.[48] By 11 o'clock, only the bare walls of the House of Lords Chamber were left standing,[49] and elsewhere the fire was blazing with brilliancy through gaps which had previously been windows.[50] The moon had taken on a curious aspect, its 'pale and silver light overpowering that of the glowing furnace which raged in the Palace'. Far from blotting out the full moon (a day away), the glow of the fire was reduced by the moonlight. As for the towers of the Abbey, they 'seemed to be sleeping in the clear moonlight, tinged also with the hue of the flames'.[51]

Just as the crisis point was reached in the Hall, with flames entering the building, 'an interference more providential than any now came'. The wind suddenly shifted more westerly, and, with the exception of the flames at the committee room corner, turned the fire riverward, carrying it away from the Hall. The flames of the burning House of Commons were pushed from the south-east gable, and the strong wind instead 'drifted the volume of smoke and kindled embers across the Thames'. From that moment on, the Hall's destruction 'could no longer be dreaded'.[52] By eleven o'clock the flames had clearly moved away from the southern end of the Hall, although Mr Solomons, one of the firemen on the leads outside the south window, was severely injured in the head by falling masonry from a building forming the southern part of the flat enclosure.[53] The engines had so successfully beaten back the fire, that, although it had consumed everything except the beams and walls of the Commons beyond, 'it had made no further impression on the Hall than by causing extensive fractures of the glass of the window'.[54] Some glass had already fallen when the lead holding it in melted with the heat.[55] All was confusion and chaos. A partial attempt was then made to save some papers from one of the nearby offices, the lower part of which the fire had not yet reached. With the distraction consequent on this, said one account, the fire was again allowed to near the Hall through a side-passage, and for a time it was again in danger.[56] At midnight, the fire caught Commons committee room 12 on the principal floor opposite the Abbey, and abutting the Court of Chancery (and thus the Hall roof again). Some guards aided by firemen under orders from both Braidwood and Mr Rooke, the foreman of the County Fire Office, set about to unroof it and cut its floorings away to prevent the fire spreading, and this 'they happily effected'.[57] The fight was not over, but the immediate peril was.

'If the windows [of the Hall] were wholly or partially destroyed, certainly not more injury was done to it', claimed one account.[58] That is not quite true. Just before the First World War, the great hammer-beam roof was found to be in a state of near-collapse due to deathwatch beetle infestation. Extensive repairs had to be undertaken between 1913 and 1922, and the roof shored up with steel trusses to prevent it from tumbling down completely.[59] For deathwatch larvae to flourish it requires the heartwood of the oak already to have been compromised by fungal decay. In historic buildings in Britain, oak beams were usually assembled green, allowing decay to be present from the very start. But, in many cases, structural damage occurs very slowly where the necessary environmental conditions are hovering around marginal thresholds—in unfavourable conditions the beetle lifecycle may be up to twelve years. A moisture content of over 14 per cent is required for a flourishing colony, and if it is under 12 per cent, the larvae will die. Even in a fairly well ventilated roof space, the normal moisture content of structural timber averages 14–15 per cent and, in many buildings in which this beetle is a problem (such as irregularly heated churches), condensation coupled with poor ventilation can significantly increase this moisture level.[60] It is therefore entirely possible that the enormous quantities of water thrown over the roof in 1834, combined with insufficient drying out, and then followed by many years of disruption to the environment in Westminster Hall caused by the changes attendant on the building of the New Palace of Westminster, tipped some minor existing fungal growth in the oak into a fully blown and catastrophic infestation by the beginning of the twentieth century. The direct effects of the fire were therefore still being felt some seventy-five years later. Even more recently, conservation work on the statues of English kings in the Hall niches in the 1990s (the figures having been unable to be exhibited at the Royal Academy's blockbuster *Age of Chivalry* exhibition in 1987) revealed that their poor condition was due to blistering and delamination of the stone. The likely cause of this flaking was identified as the repeated baking and cooling the statues had received under the south window, one hundred and fifty years before, from the heat of the nearby flames and the torrents of hot and cold water poured over them from the roof above on the night of 16 October 1834.[61]

But this was all in the future. At the time, Pugin—along with many others—rejoiced: 'I was fortunate enough to witness [it] from almost the beginning till the termination of all Danger as the Hall has been saved (which is to me almost miraculous as it was surrounded by fire).'[62] And, wrote 'C.H.' to the *Morning Chronicle*, 'amidst much calamity there is generally some cause for consolation. The destruction of the two Houses of Parliament is an event which every man who is alive to his country's fame must deeply deplore; but his regrets must be greatly chastened by the consciousness of Westminster Hall being preserved.'[63]

Milton's Pandemonium

THE CRUSH ON the Abbey side of the Palace was now so great that the police, aided by the military, were struggling to prevent the dangerously dense crowd from bursting into the spaces cleared for the salvaged furniture, books, records, paintings, and other objects evacuated from the buildings. The confusion was such that four or five people had already been run over by carts, fire engines, and carriages, and had been conveyed to hospital. Several serious accidents had also occurred to firemen, caused by the fall of burning timber or molten lead. Many of them—it was thought—were likely to need invaliding out of the service.[1] Luckily, the new Westminster Hospital had opened at the north-west end of Westminster Abbey in Broad Sanctuary earlier that year, and that is where most of the casualties went to be looked after.[2] There the most seriously injured arrived on shutters and received 'the greatest attention and care'.[3]

Of the identifiable victims of the fire, two were from the LFEE: Sub-Engineer Hambleton, whose leg was fractured by the fall of the Speaker's House roof, and Junior Fireman Solomon, who received a severe contusion on the head when part of the building outside the great south window of Westminster Hall collapsed while he was standing on the leads with the branch pipe.[4] Two other firemen, not with the LFEE, were also seriously injured. John Hamilton suffered an injury when, at 10.30 p.m., part of the outer wall of the House of Lords had fallen in with a 'tremendous crash', and he was caught by falling timbers.[5] Another fireman, Thomas Rowarth, suffered a fractured skull. Labourers and mechanics hurt included George Simmons (broken thigh when run over by an engine), John Hay (dislocated shoulder), John Slater (dislocated shoulder and hot lead burns), Michael Penning (broken arm), Thomas Gorath (severe head injuries), and Charles Boylan and Ralph Raphael (both with fractured skulls). Michael Finny ('hand and arm dreadfully lacerated') and John Slater ('dreadfully lacerated on both shoulders') appear to have been showered with hot metal or shards of glass. One woman, Rosannah McCale, broke a leg when she was run over by a carriage, and a great many minor injuries were dressed at the Hospital and the recipients discharged.[6] The King ordered a sum of money to be donated to the injured at Westminster Hospital, and the government offered compensation to those injured who were volunteers, but not to those who were hurt in the line of duty.[7]

There were many more unreported incidents, if the experience of George Manby, the fire-extinguisher inventor, is anything to go by. Diving out of the way to avoid being run over by a fire engine (how he avoided being killed or seriously injured he knew not), he badly twisted his knee in the process:

> I had just the power to reach the iron railings by the side of one of the doors entering into one of the Law Courts, when a Good Samaritan seeing my state, aided me to [New] Palace Yard opposite the door of Westminster Hall, put me into a cab and left me without having the opportunity of thanking him or enquiring his name; while in the cab before I drove off the great [north] entrance door into the Hall being

open, and the people kept from entering it by the open railing, the light reflected through the great window at the [south] end of the Hall exhibited a splendor [*sic*] awfully grand.[8]

Another person attempted to go up a heavy, unsecured ladder and it slipped back onto someone else.[9] A little boy was run over but apparently not much injured. Scuffles began to break out. A 'slight row' occurred in consequence of the police arresting a private of the Oxford Blues who was said to have taken some articles out of the wreck of the House of Commons. Nearby spectators interfered and some blows were given by the police. As a result, one youth was injured and taken away unconscious.[10] Ladies were fainting in the suffocating crush; those who helped them found their own places in the crowd immediately overrun.[11]

Sir Thomas Phillipps, the well-known antiquary and collector, was returning home from a party at eleven o'clock. He saw the blaze from a distance, and when the coachman informed him what was going on, he demanded to be driven to the spot immediately, fearing that the records of the Augmentation Office (a historic part of the Exchequer) were in danger. When he arrived he managed to pass through the bursting police cordon and 'saw what, to him was a greater vexation that the loss of the two Houses, namely a multitude of records lying in the Street over which Men, Carts, Wagons & Horses were continually tramping'. He instantly drove off to the house of Charles Purton Cooper, Secretary of the Record Commission, then living in Circus Road, St John's Wood. Cooper had just gone to bed but Phillipps roused him and waited in his coach to take him back to the Palace to see what they could do.[12]

On the river side of the buildings the flames were now wreaking the worst devastation, devouring everything in their path. The fire had early on had this third prong of attack, pushing towards the river frontage, where many committee rooms and administrative offices lay, 'all

occupied with books, papers, precedents'.[13] Also along the shoreline of
the Palace lay the Speaker's House; the east end of the House of Commons; the Clerk of the House's residence; the House of Commons
Library; and the Painted Chamber. From the Lambeth side of the
river, the previous hour's burning had transformed the view along the
river bank:

> the floors and roofs had fallen in, and the gutted buildings, glared with
> flames, ascending from the vacant area; clouds of white smoke rolled from
> the burning mass, and blue stars of fire, as it were, studded the openings
> of the windows like an illumination on a rejoicing night...Indeed the
> whole might be imagined to resemble Milton's Pandemonium; the solid
> walls, presenting numerous architectural apertures, appeared to glow red
> hot with the fervent heat.[14]

Another wrote of how the prospect on the riverside was 'a body of fire
of great length, and in some places of considerable height'.[15] It was at
eleven o'clock, presumably for just this reason, that Captain Elliot, one
of the Lords of the Admiralty, sent an express to Sir John McDonald,
Commanding Officer at the Deptford Dockyard, requisitioning the engines belonging to the Navy victualling department to provide more
firefighting capacity.[16]

The strain of overseeing the passing of the Great Reform Act had seriously undermined Speaker Manners-Sutton's health. He had wanted
to step down at the close of the 1832 session, but was persuaded by
Grey and Althorp to continue in the role despite the violent and increasingly frequent attacks of his unidentified malady. Being extremely
unwell at the end of the 1834 session, he went to the seaside at Brighton
for the summer recess for some therapeutic sea-bathing, taking with
him his daughter and lively Irish second wife whose parties in the official residence were famous—and leaving behind in the Palace his eldest
son, with a porter and two housemaids.[17] On this occasion, his wife had
decided not to take her jewellery with her to Brighton for fear of a

recent spate of cat-burglaries there, a decision she was later bitterly to regret.[18]

The Speaker's House stretched from beside the east end of St Stephen's (to which it was linked by a passage from the private apartments) along the river towards the watergate. On its west side were the cloisters. It had originally been the residence of the Auditor of the Exchequer, built around a small courtyard of the palace, but it was greatly altered and enlarged under Wyatt in 1808. His intention had been to repair or rebuild it and its immediate neighbours overlooking the Thames in an irregular, plastered, Gothick manner. The Speaker's river frontage at the north end would, he intended, resemble a 'grand, old, dwelling', with the House of Commons appearing to be its chapel. The result was (it was true) large, but rather bland and undistinguished with an insipid, crenellated exterior. The gorgeous adjoining cloister yard of St Stephen's—horrifyingly—was turned into a kitchen, with the lower oratory, originally the chapter house of the St Stephen's canons, made into a scullery.[19] No doubt this was to facilitate the use of the undercroft of the Chapel (situated on the other side of the cloister's southern wall) as the Speaker's state dining room, but, wisely, these alterations had by 1834 been reversed. The interior of the Speaker's House itself, a contemporary guidebook told its readers, 'is most exquisitely and tastefully ornamented with whatever is essential to the residence of an officer of such high rank'.[20] The state dining room itself was richly furnished with a mahogany table seating thirty-six, three black and gold japanned sideboards, four mahogany side-tables fitted round the medieval pillars, and with scarlet drapes over the windows and doors. The other formal rooms were a similar riot of red curtains and silk tassels, gold and black furniture, crimson sofas, bronze light holders, and gilt-framed mirrors.[21] It was, in fact, a palace within a Palace. Once the House of Commons caught fire, it was only a matter of time before the narrow corridors and interconnecting rooms would lead to the Speaker's House being in danger.

The upper floors of the Speaker's had begun kindling earlier in the evening, probably due to the ingress of burning smoke moving through from the Commons. In addition the wind, still continuing to blow 'very fresh' outside from the south-west, 'carried quantities of the burning materials' and set the roof alight. One paper reported that it took fire about eleven and within half an hour was almost entirely destroyed; another that it was ruined by midnight. A third reported—wrongly— that it had caught alight at 1 a.m. If the wind continued to increase, and remained in the same quarter, further fears were entertained that a range of small houses running east from the Speaker's House would be destroyed.[22] In fact, the fire was stopped in its northward tracks by a group of soldiers on the roof, who unroofed the tiles under the command of a Captain Colquohoun. They formed a temporary firebreak, but not before half the residence was devastated.[23]

Manners-Sutton's son Charles, who was about to head off in a carriage for dinner at the outbreak of the fire, seemed strangely indifferent to the danger to himself and the house, worrying mainly about the family's three services of silver.[24] He concentrated first on getting all the family's jewellery, miniatures, rings, and small *objets d'art* outside, where they were piled 'in one promiscuous heap'.[25] Many troops were drafted in to help save the residence of such a prominent public figure. Unlike at the Rickmans', where time was on their side, and the residence was considerably smaller, no one was much concerned about protecting the elegant furniture, fixtures, and fittings during the operations to remove the house's contents from the oncoming flames. Almost everything that could be removed from the Speaker's was, and scenes of chaos ensued which had to be seen to be believed. Pictures, paintings, books, 'magnificent glasses', clocks, globes, and antique desks, chairs, and tables were unceremoniously grabbed and lay scattered all over the lawn.[26] There was a mass of destruction; more or less everything was damaged by the soldiers dragging items from room to room and then onto the grass in front of the Thames, which was 'swamping with water from the engines'.[27] Several marble chimney pieces were prised off by force, a

number of doors were rammed by bulky furniture being shifted out carelessly, windows were smashed to allow items to be thrown out, and the
rooms and passages soiled by the constant tramping through of mud and
water—and, of course, by smoke damage which coated all the surfaces
with black soot.[28] Much of the furniture was thrown out of the windows,
only to be hit by cascades of water when it got outside. In addition there
were eight or ten rooms full of furniture which was not evacuated, and a
further three or four storerooms, the contents of which were mostly
burnt. In the opinion of a carpenter and builder who helped the evacuation, the house was 'very much crowded with furniture sufficient, I
should have thought, to complete two or three houses; and most of that
furniture was destroyed'.[29] The furniture in the state dining room (inaccurately known as 'Guy Fawkes cellar') seems to have survived, and,
judging by a watercolour of the aftermath, the crypt ceiling had remained intact. The former undercroft of St Stephen's was subsequently
used as a store for salvaged stonework, probably from the horribly burnt
cloister vault.[30] The valuable paintings of George III and the Duke of
Wellington about which the Speaker was particularly anxious were saved
by Mr Adamson, a policeman of the B division, who took them to the
station and then returned them in the days that followed.[31] The residence also contained at least twenty-four paintings of former Speakers,
the earliest being a portrait of Sir Thomas More, Speaker in 1523, which
were evacuated.[32] It was no doubt here that Patrick Furlong had spent all
his time getting things out as fast as he could, and generally helping with
the rescue operations, spurred on by a horrible sense of guilt. He had
become separated from Joshua Cross earlier.[33]

Looking back towards the Abbey from the Speaker's, in the midst of
the inferno, the view was breathtaking, reminiscent of the volcanic erup-
tions familiar to those who had been Grand Tourists in their youth:

> through the glowing framework of the falling walls, which, containing
> much old dry wood, continually shot forth coruscations of brilliant
> sparks, was seen an apparently interminable burning gulph, in the midst
> of which still proudly stood the walls of that Chapel [in] which had

echoed voices which had made tyrants tremble and freemen to rejoice, and which alas! for perhaps the last time had contained the wise, the good, the magnates of our land. Turning towards the Speaker's Garden was beheld the bright green sward upon which the red flames and pale moonbeams threw a varying light, covered with costly goods, strewn as chance or hasty zeal directed, while the pacing sentinel or kindly neighbour, watching over them, and varied uniforms of the soldiery (there to save and not destroy) who, emulous in exertion, were cheering each other as the fire gave way to their efforts with the engines, imparted a soothing humanity to the scene. Beyond, the Thames ran darkly on towards Westminster Bridge...covered with dense masses of human beings, the hum of whose voices might yet be heard in the more silent periods of the crackling flames; and then to look upwards and behold the silent moon, shining softly in the heavens, and with mute eloquence telling that all there was peace, filled the mind with feelings indescribable.[34]

Of the unburnt portion of the House (which must nevertheless have been severely affected by smoke and water), 'some fine relics of ecclesiastical architecture' were apparently preserved—perhaps referring to the unaffected sides of St Stephen's cloister.[35] Henry, Manners-Sutton's younger son, aged 20 (and a future colonial governor of Trinidad and then Victoria, Australia), informed his sister Charlotte two days later that,

though matters certainly are bad, they might have been much worse. A great deal of property has been damaged or lost, but still a great deal has been saved. We have reason to hope that my father's Private Library is safe. The plate and pictures all unhurt. In short the loss does not so much arise from the fire as from the removal...All the upper part of the house has been destroyed as far as my room and my Father's private library, which were the last to fall. The lower rooms are all burnt as far as the private dining room which is safe. One of the oratories and two of the galleries [of the cloister] have fallen. The rest is standing, but of course presents the most desolate appearance....*None of us* with the exception of my Father, has lost a single thing, as far as we know at present, the whole loss being in furniture and books. NB My Father, at whose elbow I write this desires his most affect. Love. He, as you may guess is up to the ears in business.[36]

Charles Manners-Sutton junior, writing from Brown's Hotel in Palace Yard the day after, was even more upbeat:

> you will hardly wonder at my not having answered your letter before, you will have seen by the papers what has happened as well as I can tell you. The fire broke out just as I was going out to dinner and dreadful it was. Most of the furniture and private property of my Father's is saved, though in a very mutilated state from removal and water. The Governor estimates his personal loss at between 2000 and 3000 pounds, all the plate we believe to be right, of course there was a great deal of stealing but out of such immense piles and heaps of things we cannot yet find what is taken and indeed the discovery must be very gradual...The house is gone from the House of Commons to the Governor's Library and on the other side to the little oratory, the rest still stands but totally unfurnished even to the chimney pieces, everything having been removed as we had not the smallest hopes of saving one single stone of the whole house, nor should we, but luckily the wind changed, what is lost and what is not we cannot yet tell...where we are to live I am sure I don't know—at present we are all over London, you will see where I am to be near my work. The Governor and my Lady are at Powell's [Hotel], she came up on Monday, Henry came up for a day [to the Carlton Club] but is gone back to Cambridge'.[37]

The boys' father, however, did not agree with their rather over-optimistic assessment of the extent of the damage to his possessions. The Speaker received an express about the fire sent to him by Charles during the night, which must have swiftly evaporated any benefit he had received from the spa treatments and ozone of Brighton, and he was expected in town the following afternoon.[38] In fact he arrived back from Brunswick Terrace the following morning, after a journey of just five hours thanks to the new Brighton-to-London 'Wonder Coach'. He was particularly anxious about his official papers. He discovered

> the state of destruction so great and so complete, as I declare I could not have calculated upon or have conceived, if I had not actually witnessed it...I could not find one single thing that existed that was not materially damaged, a great deal entirely consumed to such an extent, that the bedrooms, bedding, and furniture

of every description were all completely gone; with respect to my private library, somewhat more than half the books which I most particularly valued were destroyed, besides the book-case and every thing in the room. With respect to the other proportion of my private library, I observed that some of the books were burnt, that is, singed and mutilated, and their backs broken, and that all the rest, as I should conceive, were destroyed rather by the measures taken to put out the fire itself, for they were all thrown out into the garden, and were deluged with the dirty water from the fire-engines, covered with mud, and in some instances, where the sets were large, they were broken in by volumes being missing.[39]

Pushed beyond endurance by the devastation, his ill-health, and the hasty return from his rest-cure, Manners-Sutton was not disposed to approach the Privy Council inquiry into the causes of the fire, on which he sat, in any logical fashion. During the questioning of witnesses he tipped over the edge and 'displayed such an abundance of [folly] quite sufficient to lead to any preposterous or ridiculous opinion, and to extort from Melbourne as he went down stairs, an involuntary exclamation of "What a damnation fool that Speaker is!" '[40] Manners-Sutton eventually had to throw himself on the mercy of the Treasury to claim compensation for his belongings amounting to the huge sum of £6,000, having been unable to get insurance because of the lack of party walls in the building.[41] The fire damage was extensive, running even to the household's basic necessities:

> I had scarcely a bed existing in the house, and in order to make up the beds, a month or five weeks afterwards, that could be usable, I was forced to have the head of one bed arranged so as to fit the foot of another bed, and to manage in that sort of way. I found all my household linen which was out was destroyed; I found that, in respect to that which had been put into the chests, the family being out of town, though it was not burnt, the intensity of the heat was such, that, being doubled to be put into the chest, when taken out of the chest it tore off in squares, formed by the admeasurement of the chest, like blotting-paper; such was the intensity of the heat upon the chests.[42]

The huge crowds, piles of salvaged property, open doors, and the ability to hide among the milling groups of well-intentioned volunteers, provided too much of a temptation for some. 'Many with the sharp hyena look of plunder' soon got to work across the site.[43] 'There were vast gangs of the light-fingered gentry in attendance', noted a *Times* correspondent, 'who doubtless reaped a rich harvest, and did not fail to commit several desperate outrages, which were however much checked by the exertions of the police.'[44] The blue lobsters did deal with 'the swell mob', as well as with a number of 'suspicious and loose characters' who had got into Westminster Hall. James Leech was arrested coming down a ladder when a silver sugar basin and milk jug dropped out of his clothing. John Young and David Weldy were charged with having a blue bag stuffed with locks and brasswork from the House of Lords and four more pickpockets were arrested by constables. Another pickpocket, William Guest, was found with three snuff boxes and two silk handkerchiefs on his person, while a Francis Palmer absconded with a roll of white crêpe. Samuel Cox, a private in the 3rd Regiment Foot Guards, was remanded following a charge of stealing some flannel, an apron, and a napkin belonging to the housekeeper of the House of Lords; but his officer gave him an excellent character reference. William Atkins was charged with stealing a number of notebooks and a box containing a mirror; John Mason and Robert Taylor with the theft of a gentleman's watch.[45] The Speaker's did not escape ransacking either. Lady Manners-Sutton's jewellery was mostly destroyed, damaged, or stolen; the poor lady was still 'crying and much overcome' some days later over her losses.[46] She did manage to retrieve some valuables though. Later that week two lads, Charles Barwell and Thomas Barker, were brought before the magistrates at Hatton Garden for possession of a silver vase which they had tried to sell to a goldsmith on the Tottenham Court Road. Suspicions were aroused by its antiquity; the fact that it had been beaten quite flat; and by the accused's highly unlikely story that they had found the vase in a field while out on a fishing trip. Lady Manners-Sutton identified it as stolen from her cabinet, which had been broken into on the night of the fire.[47]

There were other allegations of bad behaviour at the Speaker's. The Guards were said to have made free with the Speaker's wine cellar instead of attending to the rescue operations; and 'had gone home quite drunk'.[48] An inquiry at the barracks on Saturday, 18 October, after the inebriation claims had been published in the papers, proved that they were unfounded and that the soldiers had been nothing other than very professional. Thomas Spring-Rice, Secretary to the Treasury, robustly quashed the rumour in a letter to *The Times*. It would be tempting to smell a cover-up in this internal investigation, but for the fact that Joseph Hume himself corroborated the story. The Radical MP lived in Bryanston Square, near present-day Marble Arch. He had been present at the fire since 7.30 p.m., finally leaving at half past midnight, and was actively occupied at the Speaker's House for more than two hours of that time, 'until all was cleared out'. In his view,

> nothing could be more exemplary than the conduct of both officers and men...soldiers and policemen vied with each other in exertions to remove, with the least damage possible, all the books, papers, furniture, pictures, and other valuables. The arrangements made by the Commissioners of Police and commanding officers for the protection of the property so removed were admirable; and so far from the least disposition to intemperance being exhibited by any of the men, they were almost exhausted by fatigue before even a very small supply of beer was obtained.[49]

While Hume's account of the removal operations at the Speaker's House does not quite accord with the evidence of the extensive damage done, it seems that the soldiers, at least, were exonerated. The upshot was that the King congratulated Lord Hill, and the royal thanks were read out to the men.[50] If anything, there was a problem not with dereliction of duty on the night, but of overzealousness. The police were a little too keen to arrest suspects, not least because in these, the early years of the Metropolitan force, it was still the responsibility of individual officers to bring prosecutions: with the concomitant advantage of being able to claim any reward for themselves. Some alleged thefts—of two silver cups from Howard's coffee-house, and of some material from a nearby shop during the chaos—which were brought before the Westminster sessions a week later were summarily dismissed. So too were the charges against John Young, a respectable Scottish baker, who was accused by Police Constable Robert Grose of stealing an engraved seal and an inkstand from one of the upper rooms in the Lords. Young explained that he had been passing at the time of the fire, and had been called in to work on the engine at the Speaker's House, and was afterwards directed to save property by Lord Melbourne himself, having been given some beer to aid his work. During the clearing of some drawers the items had simply fallen out and he had stashed them in his pockets, not intending to keep them.[51] George Phillips, a similarly respectable youth, was cleared of stealing three bodkins, and the police were roundly criticized for these and three other trivial charges by the magistrates, given the nature of the public property being hazarded at the time.[52] A George Pitts was acquitted of assaulting the Under Secretary of State while drunk.[53]

Nevertheless, Frances Rickman and her family were mightily grateful for the order imposed by the police at their end of the Palace. Members of the crowd wandering about outside their front door in New Palace Yard were pushed back, so that 'from being a multitude, [it] became clear in a large open circle, so that everything was conveyed across and Coaches only for our use, were allowed to come up'. Robert, one of their junior servants, was very active in the fire; very proud that all the police knew him and let him pass where others could not go. He became the household's reporter as to its degree of danger. Frances stationed their cook and other servants by the back door as much as possible while they were packing up, ordering them to keep it locked because she considered that they were in as much danger from looting as from fire. This was not without good reason. Indeed, their friend, Mr Apps, collared five interlopers who had sneaked into the house and it seemed to be next to impossible to prevent strangers coming in. John Rickman later gave each of his servants one pound of sugar and a pound of tea between them as a reward for their efforts. The house was full of friends and neighbours salvaging whatever they could for the family, and they all seemed to be imbued with extra strength: the delicate Henry Taylor, in particular, unexpectedly working 'like a Giant'. Time collapsed in on itself, as it does for those in the middle of catastrophes. Hours passed like minutes. The combination of fear, adrenaline, and a strange excitement at the scenes unfolding outside affected everyone. Frances worked tirelessly, surprising herself by 'comforting and scolding that night by turns—for both were necessary'.[54]

Rickman's superior, John Henry Ley, Clerk of the House of Commons—that is, its chief administrator—lived in a grace-and-favour house built in 1760 in Cotton Garden, which in 1834 formed part of the river terrace south of the east end of the House of Commons. The Clerk's house was totally destroyed during the evening. His furniture was also uninsured, but he had previously taken the view, when considering insurance, that it would be easy to get furniture out into the garden if there was need. He was proved right, and a large proportion of the moveables was removed before the fire attacked the building. His wine cellar also was saved. However, he added testily, in his later claim to the government for his other losses, 'I ought not to suffer in consequence of my house having been set on fire in the manner in which that occurrence took place, and that those whose servants produced that event [i.e. the Lords not the Commons] should not profit thereby.'[55]

The demise of the House of Commons Library was both terrible and spectacular. The House of Commons had owned a small book collection from the sixteenth century, and had appointed a Librarian from 1818; but it had only possessed a dedicated Library room from 1828. John Soane's new building, between the Painted Chamber and Mr Ley's residence, was tall and narrow with four floors (which seemed three from the outside). The main Library was on the first floor, with the committee rooms in the two floors above, and one below. Its exterior from the river front looked like a narrow Gothic tower just 23 feet (7m) wide, although in fact it was a solid block 55 feet (17m) deep, stretching back westwards towards the House of Lords Chamber. This explains why some newspaper accounts described the Library 'tower' burning. Its first floor had a handsome oriel window overlooking the Thames, lined with tall bookcases around 13 feet in height, and with elegant tables and chairs for reading and writing placed down its centre. Thomas Creevey, MP and diarist, considered it the best and most agreeable room in London. Above the Library there were originally four extra committee rooms, two of which (nos. 18 and 19, again with an oriel window over the river) were taken over by the Library in 1832, doubling its floor space and creating an Upper Library.[56]

Before eleven o'clock, the fire, 'crackling and rustling with prodigious noise', had already devoured all its interior. From the river, the building could clearly be seen alight on every level. The roof had partially fallen in, but had not yet collapsed through the floors. The rafters, however, were all blazing and, 'from the volume of flame which they vomited forth through the broken casements, great fears were entertained for the safety of the other tenements in Cotton-garden'.[57] By 11 p.m. the Library and its adjacent committee rooms were reduced to a mere shell, 'illuminated, however, from its base to its summit in the most bright and glowing tint of flame'. The two oriel windows which fronted the river 'appeared to have their frameworks fringed with innumerable sparkles of lighted gas, and, as those frameworks yielded before the violence of the fire, seem to open a clear passage right through the edifice for the destructive element'. Above the upper part of those windows a strong wooden beam which supported the upper part of the building was burning fiercely from end to end, and, as it was devoured, those watching on the shoreline and in the boats began to realize that the whole edifice was about to collapse. Soon the 'voices of the firemen were distinctly heard preaching caution, and their shapes were indistinctly seen in the lurid light, flitting about in the most dangerous situations'. Simultaneously, onlookers could hear 'the smashing of windows, the battering down of wooden partitions, and the heavy clatter of falling bricks' from other buildings along the river frontage, all evidently being knocked away to create a firebreak to stop the flames.

Two of the King's sons were observing the scene close up. Adolphus FitzClarence, younger brother of the Earl of Munster (or possibly Munster himself: sources are confused), entered the Library to encourage the workmen to persist in their efforts to save it, just as some of the rafters collapsed. He was unaware of the danger until a labourer named McCallam seized him by the collar and dragged him out as the entire ceiling fell in behind them. FitzClarence escaped uninjured, but the rafter dislocated McCallam's shoulder and he had to be taken to hospital. Another royal brother, Lord Frederick FitzClarence (or possibly Adolphus), was with a group of soldiers and policemen in the uppermost room of a turret in the western corner threatened by the flames. Also initially oblivious to the danger, 'presently their perilous condition was observed, and a fire-ladder was reared against the side of the turret'. The men descended, with FitzClarence reaching the ground last, the whole turret above them bursting into a blaze as he did so.[58] An even more garbled version of this story appeared in the *Morning Chronicle*. Lord Frederick was said to have been marooned in a burning western turret of Westminster Hall. This seems unlikely, and is probably a conflation of Duncannon's earlier escape from the Commons Chamber with the incident involving William IV's offspring in the Commons Library.[59]

At some point during that hour, the firemen gave up on the Library and left it to its fate, turning their attention to deluge the neighbouring, as yet unscathed, houses by the shore with water to protect them. 'Through a vista of flaming walls', *The Times*' correspondent wrote, 'you beheld the Abbey frowning in melancholy pride over its shattered neighbours. As far as you could judge from the river, the work of ruin was accomplished but too effectually in the Parliamentary buildings which skirt its shores.'[60] This description corresponds eerily with a painting now in the Palace of Westminster Art Collection, which, though in a naïve style, must offer a very accurate depiction of the scene from the river side: the full moon, the brisk wind, the rippling water covered with boats, the muddy shoreline packed with spectators, and the floodlit twin west towers of the Abbey.[61] At a little after midnight the front wall of the Library fell inwards with a 'dreadful crash' and shortly afterwards the flames, darting up in a startling blaze, were almost immediately quenched in a dense column of the blackest smoke.[62] In that collapse, about two-thirds of the Library's holdings were destroyed.[63]

The collections—much larger than those of the Lords Library—were regarded as invaluable by their users.[64] Nearly all the volumes in the Upper Library—about 2,000 of them—were lost. These included a very fine set of works recently presented by the French government. A blackened and battered copy of the 1830 catalogue labelled 'Upper Library' survives, which reveals what was lost. Most of the books in the lower library room were thrown out of the windows into Cotton Garden, and—like the Speaker's books—suffered a huge amount of water damage from the fire hoses.[65] Thomas Vardon, the energetic 35-year-old Librarian of the Commons, was in town, and presumably oversaw these rescue operations.[66] Although an inventory of surviving books and losses was presented to the Library committee early in 1835, it was inevitably incomplete, as not all the books in the collection had been catalogued before the fire due to the acquisition of new material from across the Commons. The most up-to-date manuscript catalogue proofs ready to be sent to the printers by Vardon were destroyed as well. That must have been a bitter blow, and would have made it harder to identify what had been burnt, greatly hindering salvage and recovery. Most of the collection of rare sixteenth- and seventeenth-century printed pamphlets from the Reformation to the time of the South Sea Bubble was destroyed, but a small portion of it, having been moved to the Speaker's Gallery, was not transferred to the Library in 1832 as the rest had been, and still survives today as the Parliamentary Collection in the Commons Library.[67] Almost certainly also within the Library at the time were the historic registers of the Church of Scotland from the Reformation to 1590. These had been borrowed by the Commons from Sion College, where they had been deposited in 1737 by Archibald Campbell, the Scottish Episcopal bishop who wished to keep them out of the hands of the Presbyterian Church of Scotland. Those registers perished alongside many other records in the Commons that night.[68]

As midnight approached, nearly six hours after the fire was first spotted, spectators on the boats saw how

> every branch and fibre of the trees which are in front of the House of Commons became clearly defined in the overpowering brilliance of the conflagration. As soon as you shot through the bridge, the whole of this melancholy spectacle stood before you. From the new pile of buildings, which are the Parliament offices, down to the end of the Speaker's House, the flames were shooting fast and furious through every window. The roof of Mr Ley's house, of the House of Commons, and of the Speaker's House had already fallen in, and as far as they were concerned it was quite evident that the conflagration had done its worst.[69]

11

A national calamity

A S THE MIDNIGHT bells tolled, Sir Thomas Phillipps the antiquary, and his friend, Mr Cooper, Secretary to the Record Commission, drew up at the Palace on a mission. Some of the most important records of the nation—the lifeblood of its history and identity, they believed—were in the direst peril. Before 1858, there was no dedicated building devoted to storing and protecting the historical documents of the government and the lawcourts. Instead, the public records from the twelfth century onwards were kept in a range of unsuitable locations spread across London. There were major accumulations at the Tower of London, the Rolls Chapel in Chancery Lane, Lincoln's Inn, the Temple, Somerset House, Carlton Ride, Whitehall Yard, and Spring Gardens. At Westminster, there were collections in the Abbey Chapter House and the Palace itself.[1] Now, because of the fire, the public records stored there were in horrible danger. So were the Parliamentary records.

Despite the removal of large numbers of public records from the sheds in Westminster Hall in 1831, those of the Augmentation Office remained in an upper room close to the Court of the Exchequer in the Law Courts. Other Exchequer records were in one of the turrets of the Hall, and, of course, there had been the tally sticks too. The Augmentation Office, a sub-department of the Exchequer, was responsible for managing all the lands confiscated by the Crown since the reign of Henry VII, and it came into its own when later charged with accounting for the royal revenues arising from the Dissolution of the Monasteries. Its records were a terrifically important historical source. Since the reforms to the Exchequer in 1833, they had lived in a kind of limbo.[2] Unlike the tallies, these were records which the government wished to keep for posterity. Phillipps and Cooper were desperate to save them.

For some hours, the crowds had watched as thousands of records were hurled out of the windows opposite St Margaret's. As they fell from the upper floors, the leaves separated and the wind blew them in every direction, causing 'irreparable injury'.[3] Bundles of documents burst as they hit the pavements and, as they scattered, all manner of curious people both respectable and less so picked them up, unaware they were committing an offence. The papers seemed to be 'of little value to anyone' in their soiled and trampled state. Only the idea that they might belong to the Court of Chancery, situated in the Law Courts, prevented one spectator from taking them into his temporary custody, because he had a 'phobia truly horrible' of Chancery, and 'of all that belongs to it'.[4]

From midnight until three or four in the morning Phillipps and Cooper endeavoured to save the scattered and perishing documents from total destruction, employing soldiers to collect them together, as well as trying to retrieve them themselves from the street outside. They were such a noticeable pair that their behaviour led later to accusations that 'some well-dressed persons' had been making off with the records, having prised them from where they lay embedded between the flooded and muddy cobbles. Phillipps had to write to *The Spectator* the following week to justify his actions. 'If these therefore are the well-dressed Gentlemen alluded to,' he informed its readers, 'the Public may rest assured that nothing has been lost through them.'[5] However, the trauma of that night stayed with the obsessive and peculiar Phillipps for the rest of his life. One obvious sign of it was his subsequent preoccupation with the danger to his own extensive antiquarian collections from fire. Visitors to his house thereafter were surprised but impressed to find piles of coffin-like boxes containing his books (with handles and drop-down lids) lining the walls. They would enable his library to be swiftly dismantled and evacuated from the building if a disaster occurred.[6]

Charles Purton Cooper was a lawyer, appointed in 1832 as the new chair of the Record Commission by Lord Duncannon. The Record Commission, the government body charged with caring for the public records, did not have a very great reputation for efficiency or effectiveness. Originally intended to oversee preservation, cataloguing, and the publication of transcripts, by 1836—when an inquiry took place into its workings—the Commission had become a byword for profligate spending, chaotic management, and out-of-control staffing. It did not have enough money to build a single, dedicated Record Office for the nation's archives, so the collections remained scattered across the city in parlous conditions.[7] Cooper himself was responsible for racking up an enormous debt of £24,000 on the publication of historic record editions over a period of just four years.[8] The Record Commission was also riven with factionalism, political back-biting, and academic jealousies.[9] The 1836 inquiry into the Commission not only uncovered accusations and counter-accusations of generalized incompetence, but also of neglect of the Augmentation records on the night of the fire.

Edward Protheroe, one of the Record Commissioners and a former MP, had been warning his colleagues for some time that the records in the Stone Building were vulnerable to fire. On the night of 16 October they were in very great danger. Protheroe happened to be passing the Palace shortly after the fire commenced. Already on the scene was William Black, a newly appointed archivist employed by the Record Commission. Although his main work was as a transcriber and editor, he lived in nearby Lambeth, and had hurried across the river as fast as he could on the outbreak of the fire. Black urged his superior to shoulder the door in to save the records. Protheroe did not believe he had the authority to do so. Everyone around him was remonstrating 'that the records must be burnt if the door was not burst open', so reluctantly he did just that. The men, with a group of soldiers, began bundling the records into baskets, boxes, sacks, and drawers—anything which could be used as a container to ship them out of the building. Henry Cole, the extremely able archivist who had dealt with the rat-infested collections from Westminster Hall the previous year, arrived between seven and eight and also pitched in. He and his new wife Marian had been sitting in their apartment in the Adelphi after dinner when the alarm was raised by a Peter Paul, one of the Record Commission workmen. Cole had shortly before been zealously cleaning and sorting the very records now at risk, his lazy predecessor Mr Caley (who had died in office) having been accused of neglecting and even vandalizing the collection. When the sacks ran out, no further receptacles could be found. Ever practical, Cole decided the best thing for it was to throw the records out onto the dry pavement. The soldiers were keeping the crowd off, so the large rolls ('not likely to suffer any injury') were dropped out of the window, ready to be sent to St Margaret's. However, the street soon became flooded with water from the fire hoses. 'Consequently,' Protheroe explained later, 'if any of the records were damaged or exposed to wet, or thrown into the gutters, the alternative of throwing them into a pool was better than leaving them to be burnt.' The threat was very real. The fire had reached the room next door and any change to the direction or speed of the wind might well have consumed the shelves of Augmentation records. Up to the last moment, Protheroe and Cole continued to work. The flames were closing in when Lord FitzClarence told them that the gas had not been let off, and was likely to explode below. They left. Out in the street, the archivists collected the remaining records from the ground and passed them through to St Margaret's Church, where they were put under guard to prevent thieving or damage. Believing no more could be done, Protheroe left first, followed by Cole and Black. When Cooper and Phillipps arrived at midnight, they discovered the archivists had gone, and that the Office, in fact, had not been burnt.[10]

At the 1836 inquiry into the Commission's effectiveness, Cooper publicly denounced Cole for allegedly not doing his duty towards the records on the night of the fire. 'It is very much to be regretted', he stated, 'that they were thrown out of the window. As I am well acquainted with the construction of adjoining buildings, if I had been present, I should have done my utmost to prevent it.' In response, Cole claimed not to have left the fire until every record had been placed in safety. All the documents of the King's Remembrancer's Office (another part of the Exchequer, some of whose records were also in the room) were in boxes, and apart from one thrown out carelessly, while he was away at the church, Cole asserted that all had been removed to St Margaret's 'in safety and without disorder'. If Cooper had been on the spot (which he was not, until the early hours, said Cole) he would have seen the impossibility of doing anything else. The gutters were dark, the streets 'half covered in water; and in removing the records, which I did, I was completely wetted through before I had finished, and up to my knees in water', protested Cole. Mr Cooper, said Cole, had misrepresented the situation: only a few documents were concealed under that water. Cole did not believe that Cooper and Phillipps could have retrieved more than one or two small rolls from the Augmentation Office. But, as he pointedly observed—being no great respecter of Cooper's technical abilities—'I know a great many of the House of Commons' papers (petitions and returns to the Medical Committee) were sent into the Augmentation Office, and produced a great deal of confusion; and it is very possible that Mr Cooper may have mistaken these modern petitions for the [ancient] documents belonging to that Office.'[11] In the end, Cooper had to back down and admit that any loss to the Augmentation Office records was minor.[12] The year after the fire, Henry Cole was sacked for blowing the whistle on the Commission. This led to the 1836 inquiry, and its series of ultra-critical findings about Cooper's incompetence and the Commission in general.[13] Employees closed ranks. Cole's erstwhile colleague William Black (a Seventh-day Adventist) detected 'the cloven foot of that devil Cole' in the inquiry's conclusions.[14] Any cooperation they may have had on the night of the fire was ended.

Nearby, in the same building, some records of the House of Commons were being rescued. An unnamed volunteer who had tried to help earlier in Westminster Hall between seven and eight then went round to the Commons Entrance.[15] There, he found many men busily engaged in carrying books out through the Members' Waiting Room on the ground floor. The voluntary salvage teams' position was 'very perilous'. Aided by soldiers, they were bringing volumes from the top of the building down the increasingly dangerous staircase, in the process being threatened with flames and drenched in torrents of hot water falling from the pipes of the engines. He joined in, carrying down many of the books, which were being placed in coaches hired for that purpose and then deposited inside St Margaret's. Once all the books were removed from that area, the flames took over. He then found himself in Commons committee room 15. This was on the second floor of the Stone Building, next to Bellamy's and facing onto St Margaret's Street. It was occupied at the time by the papers of a committee investigating the regulation of the medical profession in the room below. The committee chairman was Henry Warburton, a Radical MP who had steered the 1828 Anatomy Act through Parliament, reforming the law on the supply of bodies for dissection and thus putting an end to body-snatching. Three or four other men were already in there clearing the huge piles of papers only a few minutes away from total destruction. It emerged that they were records of all kinds: not just those of the medical committee. An argument ensued about the propriety of moving them without permission. Someone from the Record Commission who was there (perhaps Protheroe) went off to find some more coaches to transport them. The volunteer pulled down the large green curtains from the windows and with the others bundled the papers into them, tied them up and threw them out of the windows. The curtains were soon used up, so the majority of what were, presumably, Commons committee papers were thrown by handfuls into the street, then picked up and conveyed to St Margaret's Church opposite. Joseph Hume was believed to have been involved in the evacuation as well. By now it was about ten o'clock, and the unnamed rescuer rode to Cadogan Street to Warburton's residence, leaving a note informing him what had happened.[16]

Returning to the Palace, he entered another committee room—no. 13—by means of a fire ladder: that is, he got into the first floor via a window. He described what happened next. A detachment of the Blues under Lord Hill were directing the hoses of the engines against the flames, which had then

> taken full possession of the whole of the building over the passage leading immediately to the rooms. A breach was made in the ceiling of the passage, burning rafters were copiously falling upon it, and it seemed every moment as though it were about to fall. Knowing that the Railway-road committee sat next to Mr Warburton's, I proposed to two of the Blues...to go through the passage under the burning ruins and burst open the door of the last named room, in the hope of being able to save something; unfortunately before we had completely wrenched it open, the hatchet broke, but we could see sufficient to observe that everything, ceiling and all, was reduced to ashes. A strong party wall separated this room from Mr Warburton's, which was evidently a barrier against further extension of the flames in this direction.[17]

Touching the wall, they found it quite hot, a sign that the fire was raging on the other side. Attempts were being made to stop the flames by unroofing the upper rooms above. Around midnight or one in the morning the volunteer found more loose papers on the committee room floor and descended the ladder to take them to St Margaret's. When a policeman arrested him and took him to the station house, the anonymous rescuer was not impressed. 'I told the superintendent my name, and argued that my black face and wet and muddy clothes were almost sufficient to prove how hard I had worked in rescuing property,' he told

The Times, indignantly. Further argument ensued with the constable, but the superintendent eventually believed him. Undeterred, our salvage hero returned to the Palace, and helped out the Blues with the fire hoses in the first-floor committee rooms until about four in the morning, when 'they came down the ladder, I followed, burnt, bruised, and drenched with wet, as black as Diabolus, and persecuted withal. I got home about half past four in the morning; but slept soundly, knowing that I had rendered a reforming Government and the cause of reform some service, consequently the cause of humanity.'[18]

The following day he met up with Dr Somerville (army doctor and chief physician to the Chelsea Hospital; husband of the more famous Mary Somerville, statistician and mathematician).[19] The concentration of this eyewitness on matters medical and his apparent familiarity with Somerville suggests he may have been a doctor himself. He showed Somerville where the committee records were in St Margaret's, and collected some other medical papers together. These Somerville was to dry out at his house and await Warburton's return to London. Numerous committee papers were found the following morning 'damming up the gutters which were scattered in the hurry and confusion of the previous night'. Many manuscript records were saved, 'but much mutilated'. The printed committee evidence stored in a separate room perished. 'If many titled persons exhibited themselves outside the House, there were many commoners within', concluded the volunteer.[20] These committee papers do not survive today in the Parliamentary Archives. It may be that they were too badly waterlogged to be legible; or it could be that they were used by the committee, then destroyed once the Committee's report was published. Meanwhile, at the southern end of the Palace, a similar excruciating effort was being made by the fire engines, around 12.30 a.m., to save the Lords committee rooms, close by the Star and Garter public house.[21]

The account of the rescue of the documents from the committee rooms is a tantalizing glimpse of what records might have been among the archives of the Commons. For, in fact, nearly all the Commons

parchments and papers were burnt to ashes, in one of the greatest archival disasters the United Kingdom has ever known. The event is
comparable to the fire in the Cottonian Library in 1731, or the burning
of the public records in the Four Courts in Dublin during the Irish Civil
War in 1922. One of the effects of the 1834 fire was that it destroyed the
Commons' records so comprehensively, including any contemporary
indexes or location inventories, that it is only possible to find out what
historic treasures were held by the Commons through indirect (and
very imperfect) means. From a government survey which took place in
1800, it is known that at that date—officially—the Commons held the
following records: the manuscript minutes of proceedings, 1685 onwards; committee minute books since the Glorious Revolution; books
of evidence taken in the Chamber and before committees, including
those of the Committee of Privileges; and those respecting contro-
verted (contested) elections, since 1736. The mass of petitions from individuals and groups presented to the House dating from 1607 are also
gone; as are the manuscript versions of public and private bills, and
reports made from committees. Some of the latter were printed, particularly from the eighteenth century, so it is possible to reconstruct
them. The accounts and papers presented to the House from about
1607 must have been a huge mass, as their surviving equivalents in the
Lords are very extensive. The return books of the Clerk of the Crown
from 1625, the test rolls from 1698, and the qualification rolls from 1727
relating to the election of MPs and the taking of their seats: all burnt.[22]
Various other unidentified miscellaneous papers, public accounts and
unlisted boxes existed which 'appear as if they had formerly been better
arranged, as there are amongst them several lists, with the titles of
books and papers classed under distinct heads, with dates, from the
year 1495 to 1737'. This last description therefore appears to be of
the earliest Commons records, not otherwise identified by the Clerk of
the Journals. And these, of course, are only those which were known
about in 1800. They were lodged not just in the Journal Office, but in
presses in the Committee Rooms; the Long Gallery and election rooms;

in attics and even in the roof of St Stephen's. They were, the Clerk considered, 'very unsafe places for the keeping of records'.[23] He was right. Covering all manner of subjects presented, discussed, and legislated on by the Commons between 1495 and 1834, the loss of these records was a historiographical catastrophe of epic proportions.

John Rickman took a pretty brutal view of the losses, not appreciating the mass of unique records of huge historic importance which were stored in the House. He may not even have been aware of the riches squirrelled away in the obscurest parts of the Palace. 'We print so largely,' he told a friend, complacently, 'that little valuable matter is exclusively in manuscript. In fact we have gained a loss in the larger part of the papers destroyed, the mass of which [was] vast and burdensome to little purpose.'[24] No historian he, clearly. Of the losses in the Commons Library, he considered 'the good of destroying a mass of useless incumbrances is equivalent to the repairable evil of £1000–£1500 in buying books for the upper rooms of the library, the contents of the lower room little injured'.[25] Perhaps he was just being stoical, or trying to ward off the depression to which he was prone throughout his life, by putting a brave face on it. He was, however, less sanguine about his own personal papers. At the time of the fire he was posthumously editing the autobiography of his good friend Thomas Telford, the celebrated civil engineer. 'Papa was very cool', noted Frances Rickman, but 'the Wildes said he had a peculiar expression that night, he was startled most when I asked if Mr Telford's book was saved! And then sent it out directly.'[26]

Twelve men were separately charged with stealing papers from the fire as, 'during the confusion, papers of every description were thrown out at the windows; and…in their pockets were found packages of letters, petitions, and other papers, quite saturated with water. The prisoners said that they picked up the papers through curiosity, merely to read at their leisure. They thought it was no harm, as others were doing it…some of them are highly respectable tradesmen.'[27] Many others would have left the scene with records without being caught. George

Manby did, and sent one as a souvenir to a friend. Its charred remains are preserved today among his papers.[28] On 27 January 1844, the British Museum purchased at a book sale an atlas folio containing what were described as 'historical and parliamentary documents during that interesting period of English History 1640 to 1660', belonging to a Joseph Lilly. Bound inside were numerous petitions of various shapes and sizes, on parchment and paper.[29] Some were obviously water-damaged, but not all. There was also a little evidence of charring on them. They included what is today known as the 'Root and Branch Petition' of 1640. This was a call to the Commons from 15,000 Londoners for the abolition of episcopy in England, that is, the abolition of the Anglican Church hierarchy. It led to the Root and Branch bill of 1641. Also in the volume were lists of persons captured after certain Civil War battles, the commission for Thomas Fairfax to be Commander in Chief of the Parliamentary forces, and the oath sworn by Oliver Cromwell on becoming Lord Protector. It is fairly conclusive that this volume contains records which had been presented to the House of Commons, then stored, thrown out of the windows nearly two hundred years later at the fire, charred, soaked, picked up by a passer-by (maybe Joseph Lilly himself), dried out, carefully flattened and bound, and which then came onto the antiquarian market nearly ten years later. Since Parliament at the time seemed to care little for its historic records, this appears to be the best thing that could have happened to them.

There was one piece of good news. At some point, the original manuscript Journals of the Commons—231 parchment bound volumes from 1547 onwards—were saved. There is no evidence of how this happened. They were presumably held in the Journal Office on the second floor, yet the 1835 report on the losses sustained by the Library stated that 'everything in the office of the Clerk of the Journals perished in the same conflagration'. So it remains a mystery how, when, and by whom they were rescued. Today, they bear little sign of having sustained damage from being thrown hastily down stairs or out of windows, and have no scorchmarks. This would point to an orderly rescue of these, the most important core records of the House of Commons' daily proceedings back to the reign of Edward VI, in a complete run, as described in the inventory of 1800.[30] 'The loss of records sustained is not important, nearly everything of value have been printed', wrote the *Gentleman's Magazine*, in a direct echo of Rickman. Those who wish to use the pre-1834 Commons archives today would disagree. But it did at least regret 'the test and qualification rolls, signed by the Members after taking the oaths'.[31]

Cooper, Phillipps, Black, and Cole were not the only archivists to be wringing their hands that night. Across the road on the roof at the east end of Westminster Abbey, Francis Palgrave, the recently appointed Keeper of Records in the Chapter House, was keeping anxious watch.

Any change of wind direction or a stray spark threatened 'not one or more buildings of great importance to the community at large, but a whole district of the metropolis... and not least, that stupendous monument of the taste, the magnificence, and the piety of our remote ancestors, the mausoleum of a hundred kings'.[32] It seemed a very real possibility, as 'occasionally the wind, shifting on uncertain points, rolled [the fire] back, then brought it again dashing forward towards the Abbey, while, in returning, caught as it were among the inimitable fretwork and tracery of the Chapel of Henry the Seventh, it lingered an instant with a sort of playful pause before the wind dashed it again upon its office of destruction. The Abbey looked for some instants over the scene in dark and frowning quiet, but soon to its highest pinnacles the wayward light played over it.'[33]

Government records had been kept at the Chapter House since the 1540s. They moved in probably around the same time that the Commons moved out of its usual meeting place in the Abbey and into St Stephen's, across the road, at the invitation of Edward VI. Over the centuries, the Chapter House repository had been managed by a series of highly competent archivists, including the great figures Arthur

Agarde and Peter le Neve, but the condition of the building was less than desirable. As well as frequently needing repair, the Chapter House was a constant source of worry to the guardians of that portion of the public records, since the lit coppers of a nearby brewhouse and wash-house, and an adjacent coal store, posed an ever-present threat of fire. Nevertheless, the Chapter House repository had managed to survive all those dangers, and in 1834 it contained some of the most important records in the kingdom: again, belonging to the Exchequer. Alongside the staple sources used today by medieval historians (deeds, treaties, writs, inquisitions post mortem; plus the pipe, fine, patent, and close rolls; wardrobe and household accounts; and records relating to Scot-tish and French possessions) were the *Valor Ecclesiasticus* (Henry VIII's 1535 inventory of Church possessions at the Reformation), and Domes-day Book itself.[34] It was quite a warehouse.

Alongside Francis Palgrave on the leads was the Dean of Westmin-ster. The Abbey's stonework was intensely red from the light; its col-umns, pinnacles, and buttresses appearing to be built of 'fixed fire' and 'stamped in living flame'.[35] Dean Ireland, aged 72, was a somewhat conservative, unimaginative churchman with his feet planted firmly in the eighteenth century. Palgrave, on the other hand, was a brilliant and dynamic lawyer-cum-archivist, the 46-year-old, self-educated son of a Jewish stockbroker called Cohen, who had taken his wife's family name and Anglican religion on their marriage. The risk was very evident to bystanders. The House of Commons committee rooms at the southern end of the Law Courts complex had been blazing since eight o'clock and, when viewed from New Palace Yard, an alarming 'mass of fire was seen rearing its forked crests into the sky at a short distance on the left of the venerable Abbey'.[36]

The two men had climbed up onto the slates through a hole in the roof to watch. When a sudden gust blew the flames for a time across the narrow gap in Old Palace Yard towards the Abbey, Palgrave had urged the Dean to descend with him so they could grab Domesday—and the other most precious items—and move them into the nave of the Abbey for safety. Ireland, with all the phlegmatic stubbornness for which was famous, replied that he could not even think of doing so without permission from the Prime Minister.[37] We do not need to imagine what went through Palgrave's mind in the hours which followed this exchange. For the rest of his life, he remained extremely anxious about the safety of the Chapter House treasures. In fact, like Thomas Phillipps, it had a lasting effect on his peace of mind. It was a constant burden of worry. 'If any of the materials', he wrote of the Chapter House, 'were once kindled, the flames would rage with great intensity and communicate with every part, all being connected with wooden floors and wooden staircases...the whole interior would be a mass of ruins [and] it is most probable the whole would collapse.'[38] But at least on that night, the Chapter House remained unaffected. 'What a dreadful fire...How fortunate the Abbey has escaped', Lady Georgiana Pratt later wrote to her brother, the second Marquess Camden, in relief.[39]

Dean Ireland, however, was wrong to suppose that Melbourne would have been much engaged with their worries. The three fire engines stationed at the south end of the Palace complex by Braidwood, playing water over the House of Lords Library, may well have been responsible for saving what was then—and still is—one of the most famous records in Britain. The Death Warrant of Charles I, signed and sealed by the fifty-nine regicides (including Cromwell) who ordered his execution following a trial in Westminster Hall in 1649, was reported missing in the days after the fire. For years it had been stored in the Jewel Tower opposite Old Palace Yard, with the Acts, and was well known from the printed facsimiles which had circulated widely since the eighteenth century. But when a new Library for the Lords was completed in 1827, the Death Warrant was framed and glazed, and hung there as a curiosity. Some newspapers declared it destroyed; others believed it to be 'somewhere carefully locked up...although it cannot now be found, or remembered where it may have been placed'.[40] Unknown to the papers, however, the Death Warrant had been swiftly rescued on the night. 'I was with Melbourne,' wrote Broughton, 'who was as usual very cool, and now and then inclined to be jocose. He could not help laughing when a man ran up to me and said, "Sir John, we have saved King Charles's warrant!" meaning the original death-warrant, as if that document was particularly interesting to me.'[41]

The Lords Library, 'a modern, large and beautiful building', had been designed by John Soane. It was typically Soanic, and complimented the new Commons Library and the ancient Painted Chamber.[42] One account tells how it was soon completely destroyed by the fire, the roof falling in with an immense crash, although almost all the rare and valuable books and collections were saved. The rapid engulfing of the Library, pronounced one report, 'is only to be accounted for by the circumstance that the flame had little impediment in this direction, although the distance was so great, there being the great Chamber, constituting the whole House of Lords, the Robing-room and a large lobby'. Yet this story of its demise must be wrong, because the Lords Library was fit to be used a week later for the prorogation of Parliament. The official report of the destruction from the Office of Works confirmed that it was in fact undamaged; and most of its contents had been evacuated over the course of the evening.[43] What seems to have happened in this account is that the fate of the Commons Library has been mixed up with that of the Lords, where events had taken a very different turn.

Like Vardon in the Commons, the Lords Librarian John Frederick Leary was in town that night. So were most of the Lords' clerks. Together, they 'afforded very prompt aid' in the rescue effort. The most extraordinary endeavour was made to save property and books.[44] Many of the Library's contents had been decanted into Lords committee rooms during enlargement works on the Library over the late summer of 1834. In fact, some rumours circulating afterwards said it was the carelessness of a builder's plumber there which had caused the fire in the first place.[45] This earlier removal helped the book salvage teams in their later efforts; Leary, however, did lose his private collection of art in the fire.[46] A combination of wind direction and luck, the efforts of

Braidwood's engines stationed at the Library, and the thick stone walls of the Painted Chamber meant that the fire was halted before most of Soane's Library and the new ceremonial route through the Lords was damaged. A sixteenth-century Bible used by the Lords for the swearing-in of new members at the Table of the House was saved and survives, severely singed but intact, in the Lords Library today. J. F. Leary, however, was understandably nervous in the days which followed. He wrote to Duncannon, one day short of a month after the fire, still in a hyper-alert state, saying:

> I have confident hope that the precautions taken by Sir John C Hobhouse when I privately communicated the fact of having found *burning embers* beneath a grate within 2 yards of the Library will have due effect: and that the most vigilant watchfulness will be continued to be kept upon the approaches of the Library. I have nightly perambulated these approaches, I shall continue to do so; my sleeping apartment is likewise known to those who may be on duty during the night. I trust your Lordship will rest assured that no personal care and attention shall be wanting on my part. My immediate responsibility is however confined to the Collection, in fact to the interior of the Room and your Lordship will readily conclude that I have there enough to do with my deranged collection and indexes. I lay these points before your Lordship impressed with the importance of the Collection and in the hope that with due precaution no further accident will ensue.[47]

Meanwhile, the archives of the Lords were stored in two main locations. First, there was the Jewel Tower. This L-shaped stone building, surrounded by a moat, was detached from the old Palace precincts in 1834, as it still is today, positioned close to the east end of Westminster Abbey and next door to 6/7 Old Palace Yard, a Georgian residence for the Clerk of the Parliaments, the official custodian of the Lords' archives. It was the one part of the medieval Privy Palace still surviving. Probably built in late 1365 by Henry Yevele, it originally stood at the south-western corner of the Palace's walled enclosure.[48] Since the sixteenth century, the Acts, Journals, minute books, and sessional papers of the Lords had all been stored methodically inside it, on shelves and in cupboards. Secondly, the petitions, orders, returns, and bills of the House of Lords had been moved into a fireproof repository under and around Soane's Royal Gallery in 1827, as by that date the Jewel Tower was fit to burst. Among these thousands of bundles were nestling great historic and constitutional treasures, including the coded letters of Charles I captured after the Battle of Naseby, the Declaration of Breda (in which Charles II announced his intention to return to England to restore the monarchy in 1660), the draft Declaration and final Bill of Rights, a host of anti-Slavery petitions, and the minutes of the trial of Warren Hastings. Other accounts and reports remained scattered in the Painted Chamber and one of the committee rooms.[49] The Acts of the 1833–4 session, meanwhile, were still lying in racks in the main administrative office in the Lords, known as the Parliament Office, until the printers' copies could be checked against them, before their transfer across the road to the Jewel Tower. They only narrowly escaped destruction by a deluge of engine-water.[50]

While efforts in the Commons, both in the committee rooms opposite the Abbey and in the Library, were desperate, improvised, and involved bundles jettisoned from the windows, the Lords evacuation seems to have avoided this and to have been much more methodical. Rescuers of Parliament's collections at the Lords end no doubt had good luck compared with their Commons counterparts, but it is also true to say that their efforts were greatly aided by the quietly heroic efforts of Henry Stone Smith (1795–1881), Clerk of Committees in the Lords, who rose spectacularly to the occasion. He lived in nearby Queen's Square (now Queen Anne's Gate), and having 'some traditional knowledge' of the importance of the historic Lords records, although not a detailed appreciation, took charge of an enormous logistical exercise to save and relocate them.[51] Stone Smith certainly knew the locations of the collections to be saved. We cannot tell for certain how the Library's books in the committee rooms and the Lords records near the Royal Gallery were evacuated from the buildings, but it was probably done by a human chain of Lords staff and soldiers passing items down the line, piling them into barrows, carts, and other vehicles at the nearest exit, or shifting them into the gardens. Mr Aldin, doorkeeper to the Library, played a part—he was later recommended for a tip of £10.[52] The books and records were then driven off, either to St Margaret's Church, to the houses of clerks living close by, or even to the Westminster Hospital. By the end of the night, St Margaret's was 'literally crammed' with papers, furniture, and boxes of every description. Mr Forty, the churchwarden, superintended the property saved, and sat up the whole night to prevent plunder, along with soldiers guarding the hoard. Many Lords records were also removed from the Parliament Office (Soane's riverside suite of rooms for the Lords administration), but the more ancient ones remained undisturbed in their fireproof vaults ('very near the spot where the Guy Faux combustibles were discovered', *The Examiner* could not resist adding).[53] Stone Smith paid forty-six labourers for help that night; the soldiers would have done it under command; and the clerks for free. Each labourer got five shillings.[54] Stone Smith was later set to work superintending the rearrangement and checking of the Acts, prior to their removal into the Victoria Tower of the new Palace. In 1874 he was still in the employ of the House of Lords. Aged nearly 80, he lived long enough to aid the first inspection by the Historic Manuscripts Commission which made the importance of Parliamentary record holdings available to the world for the first time.[55] It was an achievement of which he could have been justly proud.

A few onlookers seemed confused about how the law worked. With Parliament gone, and the Commons records burnt, wouldn't that mean that laws were no longer in force?

Some bawled aloud good-lack-a-daisey,
The Agoney Bill must be uneasy,

190

Some said the Beer bill had stop his breath, sir.
And the Reform bill died a dreadful death, sir
Some Ancient Bills were sadly failing,
The Sweeps Bill too was deep bewailing.
And many swore when home returning,
They saw the Poor Law Bill a-burning.[56]

One tale told how a group in front of a Lambeth boathouse immediately opposite the House of Commons included a chimney sweep looking intently at the fire. A waterman's apprentice clapped the sweep upon the shoulder and asked (referring to the restrictions of the 1834 Sweeps Act): 'Well Snowball, aren't you glad of the fire; if both houses are burnt mustn't your gagging act be burnt along with it, and can't you now cry "Sweep" and "Soot-oh" in spite of Parliament?' The sweep replied that the master still had a copy at home, but the apprentice said, 'You don't mean to say he'll be such a fool as to let the Parliament chaps know that?' A similar anecdote claimed that a sweep was heard to call out lustily, in high glee, 'Ah, they'll let us cry sweep again now, I'll bet a guinea.'[57] Another story recounted how a ragged-looking man observing the removal of books and papers from the House of Commons Library earnestly asked everybody that passed, 'Whether the Poor Law bill was burnt?' At length someone humoured him by saying the bill had been saved, to which he replied, 'Worse luck to them that saved it and I hope which them as made it and them as saved it was burnt themselves.'[58] Unknown to them (not that it would have made any difference), the Jewel Tower records, including all Acts of Parliament from 1497 onwards and the Lords Journals, had remained safe and snug in that location for the duration of the crisis.

After it was all over, it was possible for one newspaper to state: 'as impartial reporters we must say that the great body of spectators did not evince much veneration for the old edifice, or in any way lament its destruction'.[59] But in contrast to this mood, another reported quite the opposite reaction. What the *Ipswich Journal* lacked in accuracy, it made up for in passion:

This disastrous occurrence—the entire destruction by fire of the two Houses of Parliament—Lords and Commons, with the library, archives, votes, rolls of the former, and there is too much reason to fear, a considerable portion of the valuable records of the latter can be regarded in no other light than as a national calamity... [the fire seems] to have destined all that was venerable and valuable for evidence or association in the history of our country to one common doom, for, not content with destroying the muniments of legislative lore, hitherto through a long series of trials, and amid a multitude of accidents, respected, it endangered the evidences of social rights, for even Westminster Hall with its large store of deposits, the all-important guarantees of so much of the property of the kingdom, was more than once within the jaws of demolition, and but saved, after all, by little short of a miracle... The records of the courts of law had been thrown out of the windows, and many of them are preserved, but all was in confusion up to the latest moment.[60]

12

Friday, 17 October 1834   1 a.m.

Emptying the Thames

IN THE EARLY hours, a sensational rumour began to circulate.
Newspapers with the first reports of the fire had been printed and
were already winging their way across the provinces: the *Morning
News* arrived in Bristol via Cooper's Coach in the early hours of 17 October.[1] It began to be whispered that at midnight a bundle of matches had
been found under a tree in the Speaker's garden. Mr Jones, a doctor of
Carlisle Street, Soho Square, had reported his find to a nearby policeman (Constable Farrell, no. 48L) and a sentry from the second battalion
of the Grenadier Guards.[2] A war of words subsequently broke out in the
newspapers in the days that followed, regarding the allegations of arson.
'Suspicion', wrote the *Bristol Mercury* of the Houses of Parliament, 'has
not scrupled to ascribe their fate to the deliberate act of some cold-
blooded incendiary (though believed to be an accident).' It was too much
the habit of the press generally, went on the same editorial, 'to add, in all
cases where the cause of the accident is not apparent, the words—*without
doubt this was the act of an incendiary*', but 'that man does not exist who could
apply the torch of the incendiary to any portion of the venerable pile of

Westminster Hall...whatever may be the feeling towards Parliament...there is no partizan who has any interest in the destruction of the place of its meeting or in what would be far worse, the placing of the public records in jeopardy'.[3] The *Morning Post* did not credit the rumours: it would be a 'revolting conclusion' to believe that such an 'act of atrocious wickedness' could even be possible.[4] The staunchly Whig *Morning Chronicle* declared that 'too much praise cannot be given to all the members of the government who were in town and who promptly afforded every aid which their presence and personal exertions could give'. The Tory press had, it believed, been too much inclined to blame incendiaries and to malign the conduct of the people but, 'by common consent, there never was an occasion in which all classes exhibited more commendable feeling, or rendered more cheerful and ready aid'.[5] Thomas Walsh, a gentleman, wrote to the Home Secretary in early November claiming he had the matches in his possession, but now wished to come forward in order to dismiss the regrettable impression abroad that the fire was the result of incendiarism.[6] Duncannon's response is not known to this, or a number of other crank letters (including one about a penny Almanack showing the Houses of Parliament *on fire*, published three days before the catastrophe, but now mysteriously destroyed as waste paper).[7]

By the end of October, the Privy Council inquiry had already heard all the evidence of Cross, Furlong, Weobley, and Wright, and its sessions had been widely reported. Yet less scrupulous rags like the *Herald* were still managing to titillate their readers by pointing out 'how anxious that Government papers are to make the public believe that the fire was purely *accidental*—that the fire originated in *accident*; but we want to see the *proofs*'.[8] However, Henry Manners-Sutton had already reassured his sister on the matter on 18 October:

> The fire still occasionally blazes up, but I believe there is no fear that it should again do any damage. You must not believe all the stories that you hear about finding bundles of matches etc in the garden, for although it has not been proved not to be the act of an incendiary, still as yet there is nothing to lead anyone to suppose such to be the case.[9]

The reason for the *Herald*'s and others' allegations was simple: it was fantastic copy. For in 1834, these stories would most likely have immediately conjured up one name in the minds of those who heard them: Captain Swing—and a shiver might have gone down their spines. The so-called Swing Riots of 1830–1 had shaken people severely. Taking their name from the call made by the leaders of haymaking teams, these incidents of agrarian unrest had been a protest mainly against the introduction of threshing machines; plus the result of distress about low wages, poor working conditions, and changes to the administration of the old Poor Law. Portable, horse-drawn threshers had begun to be introduced during the Napoleonic wars as a response to a shortage of agricultural labour. They had proved to be much more effective and cheaper than the old system of threshing, which involved the manual beating of grain with a flail by agricultural workers in order to release the chaff. The problem was that farm hands had little other employment during the winter months and, deprived of this traditional activity and the bonuses it paid, destitution soon followed.[10]

The Swing Riots were characterized by arson and machine-breaking, but also robbery, wage and food riots, wilful damage, animal maiming, and assault. At least 3,282 incidents associated with the phenomenon have been identified throughout England, Scotland, and Wales. Most occurred in the South-East and East Anglia. The victims were mainly farmers, who often received threatening letters signed 'Captain Swing'. In the words of one popular song after the fire:

> Some said that Swing had come to London
> Some said twas an accident, others treason,
> Some said indeed twas a dreadful loss, sir
> Some said they saw at Charing Cross, sir
> Guy Faux a looking out of the window.
> With a lanthon, candle, matches and tinder
> But I dont believe upon my soul sir.
> Guy Faux was near the fire, at all sir.[11]

If Captain Swing was not on people's minds watching the fire, then another identifiable arsonist might well have been. Some in the crowd began to make connections with the burning of York Minster five years previously. Those who had seen that most notorious act of arson remembered it as being a small fire in comparison to that at Westminster.[12]

On Sunday, 1 February 1829, a disturbed Methodist called Jonathan Martin attended evensong in York Minster. He had been irritated by the sounds coming from the great organ during the afternoon, and at the service itself had felt that the prayers and singing were not enthusiastic or sincere enough. The organ, he decided, was deceiving the congregation with its noise: 'I'll have thee down tonight,' he thought to himself. 'Thou shalt buzz no more!' And so it turned out. Martin hid in the north transept of the Minster after the service, and once the bellringers had left the great church and it was locked up for the night he took to accomplishing his 'great work'. Intending to burn down the organ, he broke into the choir and set about piling up prayer and music books in two heaps. He then took one of the lit candles from the lectern and put a flame to his bonfire. Escaping through an aisle window after eight and a half hours inside, he later declared, 'I had a hard night's work but the Lord helped me.'[13] The Minster suffered terrible damage.

Although Martin had fled the scene, his eccentric behaviour was hard to disguise and he was arrested within a week at Hexham. Appearing before the Assizes at the end of March, a plea of not guilty was entered: 'It was not me, my Lord, but my God did it.' Not all witnesses appearing at his trial believed him to be mad but enough did, and his defence team—including Henry Brougham—saw that he was found not guilty by reason of 'monomania'. At the end of April he was on his way to Bedlam, more properly known as the Criminal Lunatic Asylum, in London. On his journey he announced he intended to burn down St Paul's after having an audience with the King. On discovering that there had been a recent 'arson' attempt on Westminster Abbey and that God had therefore seen fit to appoint another to the divine task, overlooking his abilities and good service, he expressed his extreme irritation.[14] However, on the night of 16 October 1834 the

196

country's most notorious arsonist was safely locked up in Bedlam, just over half a mile away on the other side of the Thames. Perhaps he could see the glow through a cell window.

In the early hours of the morning the scene from Westminster Bridge was still 'awfully grand'. The Commons Library, the Painted Chamber, Mr Ley's residence, the two Houses of Parliament, the Speaker's House, and even Westminster Hall—in fact, almost the entire Palace— seemed 'one body of fire'.[15] The crowds of boats still remained, the streets still thronged, and the fire continued burning just as extensively, but perhaps not quite so ferociously, as before.[16] The two LFEE engines tackling the blaze by the river were being backed up with the two belonging to St Margaret's parish and the St Martin's engine—all based in the Speaker's Garden. They were fighting a losing battle. At 1.20 a.m. the bow front of the Commons Library wall fell forwards into the garden with a loud crash. There were unconfirmed reports that a fireman and soldier were buried in the ruins. Ten minutes later, the roof of the southern wing of the Speaker's House was taken off by a party of men from the Office of Works under Lord Hill's direction, and the hose of an engine carried through. At half past one, the naval engines from the Deptford dockyard, which had been summoned ninety min-utes before, arrived with thirty-three marines to work them, and a few minutes later four more engines pulled up, drawn by horses, with a party from the dock police, all commanded by Captain Brown, RN.[17]

At last the tide rose and the great Floating Engine, which Braidwood had summoned nearly five hours earlier, finally chugged under West-minster Bridge. It still had the name of its former owner, the Sun Fire Insurance Co., painted on its sides. One of the *Times* correspondents passed it as he returned home by water, to rest. 'It might have been of great service had it arrived earlier,' he noted grimly, 'but the state of the tide and the shallowness of the water prevented the steamer from coming sooner up the river.' Braidwood himself admitted 'it could not be brought up or got to work so promptly as under other circumstances it might have been'. In an outburst of professional jealousy, Palmer, the

St Margaret's parish engine keeper, claimed that the Float did not work till 2 a.m., and then was not wanted. Another account says it took up to an hour to get going; but a third that it began work quickly. Once it did get set up—throwing a ton of water a minute at the flames—the effect was 'positively prodigious'. By about three o'clock the fire at the Speaker's House was under control. Throughout the rest of the night the ruins continued to burn with great fury, but the other land engines remained constantly at work, pumping a stream of 2,000 gallons of water a minute on them.[18] With the unroofing which had taken place earlier in the evening, and the decisive action taken by the Floating Engine, the fire's progress at the Speaker's House was finally halted and half of it was considered saved, though the surviving fabric and contents had, of course, sustained huge damage.[19]

The barge on which the Floating Engine was fixed had been towed up from its station at Rotherhithe by the *Fly*, a steam packet with seven crew. The owners of the *Fly* were paid £10 for their trouble on that night. Someone suggested that Frank Rice, the master, and Bill Morgan, the engineer, ought to be tipped £1 each, plus ten shillings each for the rest of the crew, but this was disallowed by the Office of Woods and Forests: presumably they imagined the owners would show their appreciation accordingly from the amount already received.[20] One person, however, did not take the Office's refusal to show its appreciation of the Float's work lying down. David Bromer, of 7 Hunter Street near Brunswick Square, claimed to have 'exerted himself greatly' on the night. It is not clear exactly what he did—perhaps he got in a carriage on hearing Braidwood's call and went to Rotherhithe himself to set things up; perhaps he paid the *Fly*'s owners in advance for towing the engine. Another possibility is that he paid the many men who pumped the engine (when Braidwood later designed his own Floating Engine, it required 120 men to work it).[21] Interestingly, the entry in the LFEE's official log book suggests that there was a second floating engine at the fire: one not associated with the Establishment.[22] Nothing more is known about this engine, and the newspapers speak only of one being present. Maybe Mr Bromer somehow summoned up a second barge from somewhere. The matter remains a mystery. At any rate, he shelled out the princely sum of £12 13s. 6d. in disbursements on the night relating to 'obtaining the floating engine'. In the months that followed, Bromer worked himself up into a lather about his contribution. Expecting to be interviewed by the highest in the land about his part in the saving of the Palace, he got in touch with his solicitor, Mr Lefroy, indicating he expected some sort of personal commendation from the Prime Minister. Lefroy then racked up a further bill of £11 3s. 6d. in attending to the demands of his client, for which Bromer decided to charge the government. On 12 December, he received a brief letter from the Office of Woods requesting a stamped receipt for the sum he had paid out on the night of the fire, which would enable them to send the payment through. The expected letter of thanks from the Prime Minister was not enclosed. Bromer was furious at the snub:

> It certainly does not appear to me that my exertions were worth more than what I cannot help terming a Letter of Insult rather than that of an acknowledgement. I do not intend sending a Stamp receipt for the amount... It is sufficient for me to know I will not let the matter rest where it is—as I shall consider it my duty to inform the Public that exertions made on behalf of the government are acknowledged by the Government at all events one branch of it, in such a way as to entail considerable inconvenience & some expence on those, who may have been induced to act.

Poor Mr Bromer. He was eventually paid in full for his efforts, but never managed to persuade the Office to send through the second instalment relating to Lefroy's fees. He never received his letter of thanks from Lord Melbourne either. All went quiet for over twenty years, then as old age approached, the memory of this gross injustice returned to prey on his mind. In 1855, having attempted (unsuccessfully) to drum up support from his former Rugby schoolmates to obtain a petition to the Crown on the subject, he wrote again to the Office to request Lord Melbourne's vote of thanks be sent to him, which he was convinced had got lost somewhere in the system. When he again wrote in 1863, claiming to have 'lost' the imaginary vote of thanks and needing another copy, he left the civil servants of the Office in complete puzzlement. 'I do consider the Crown may ultimately render me the simple justice I deserve,' he wrote plaintively in a final blast, 'having been the Principal Means of saving Westminster Hall and the Speaker's House, not a drop of water *for hours* would have reached the Speaker's House and the water side portion of the conflagration but *for my Pain* and personal executions.'[23]

The Float was too late to save the Painted Chamber. On this subject, even Pugin abandoned his otherwise satirical tone. What was most to be regretted of the whole event, he thought, 'is the Painted Chamber, the curious paintings of which I believe are totally destroyed'.[24] The *Gentleman's Magazine* ranked its destruction alongside St Stephen's and the Armada Tapestries as the top three losses in the fire, for 'the lover of antient art'.[25] It is not clear when the Painted Chamber first caught fire. It was adjacent to the House of Lords Chamber, but the ancient masonry provided some protection between the two buildings. One possibility is that the flames first got hold by connecting across the roofs quite early on: perhaps as early as ten past seven.[26] Inside the Palace it was said that the Robing Room in the Lords (between the two spaces) did not succumb till one in the morning, despite being surrounded by flame. Nevertheless, it seemed to onlookers that between half past midnight and one o'clock, the Painted Chamber was 'completely reduced to ashes'.[27] The roof and ceilings had fallen in, and the walls had been stripped bare by the intense heat. But in fact the thick stone structure of Henry III's state bedchamber, which had stood for nearly six hundred years, was made of sturdier stuff, and the walls themselves stayed upright, though battered. Unlike the soaring chapel of St Stephen's, its squat shape and relatively unaltered interior seem to have protected it from total structural compromise.

In the 1260s, on restoration after another fire, the Painted Chamber's ceiling was covered with pictures of seraphs and prophets in a highly sophisticated style. Then, almost immediately afterwards (probably because Henry III changed his mind), this decorative scheme was altered and the ceiling was boarded over with planks of oak, studded with decorative bosses known as *paterae*.[28] These bosses can still clearly be seen in a watercolour of the Painted Chamber executed by the antiquarian William Capon in 1799.[29] Around 1795, John Soane, recognizing the space for what it was, had proposed restoring the Painted Chamber to its ancient state, but his plan was blocked, and the room was left to deteriorate further.[30] The ceiling survived until around 1816. It was repaired in that year, and Adam Lee, the Clerk of Works, took home four of the painted panels, rediscovered under the *paterae*, as souvenirs. Two of these survive today at the British Museum. The woodwork was still in a poor state two years later, when John Soane again surveyed it, and he himself removed one of the *patera* to decorate a Gothick 'Monk's Parlour' at his house, now Sir John Soane's Museum in Lincoln's Inn Fields. The whole ceiling was finally dismantled 1819, when it was replaced with a plaster replica.[31]

But it was the walls of the Painted Chamber which stunned thirteenth-century contemporaries and later generations. They were covered with painted and gilded murals of various angelic Virtues triumphing over monstrous Vices, and with Old Testament scenes of good kingship, the story of Judas Maccabeus intended as an exhortation to the monarchs who woke every morning beside him when lodged at Westminster.[32] By 1834, its elegant paintings were mostly invisible, obscured by the neglect and ignorance of three centuries. The fabulous wall paintings had been kept in good condition by Henry III and Edward I's successors throughout the fourteenth and fifteenth centuries. But after the 1512 fire and the displacement of the royal family, they gradually fell into disrepair. Unfashionable and decaying, they were whitewashed over and forgotten, then continually covered with more layers of whitewash and blue paper until rediscovered just a few years before the fire in 1818–19. The antiquary and draughtsman William Capon washed off some paint, and began to reveal figures of armed female Virtues and Vices in the window splays. In the summer of 1819, a full-scale operation took place to uncover the numerous paintings of 'exquisite beauty and freshness' under the supervision of another sympathetic antiquary, Edward Crocker. It was found that across the walls were bands and bands of brightly coloured narrative paintings and inscriptions, and a full-scale mural of the coronation of St Edward the Confessor was discovered on the north wall. At least three separate artists took down copies of the paintings before the sadly neglected walls were replastered. The paintings of the female figures of *Largesce* (Generosity) and *Covoitise* (Covetousness) on the south side window splays remained exposed (and can clearly be seen in a contemporary print). They all still survived at the time of the fire.[33]

The Painted Chamber was more than just a lovely medieval building. It had been one of the earliest meeting places of the House of Commons. In offering this glamorous, elegant space in his principal royal palace to the newly empowered burgesses and knights of the shire, Edward III (1327–77) was signalling his personal opinion of their worth to him. In 1340–1, there had been a serious political crisis. The King had run out of money to fund war with France, and had to return home in a furious temper to convince Parliament to raise more taxes to pay off his debts, and enable campaigns against the French to continue. The Commons had first been summoned to Parliament to represent the towns and shires in the thirteenth century. Boosted by the economic and political crisis of 1340–1, the Commons first met separately from the Lords, and flexed their collective political muscle to gain favours from the King. It was a cunning move by the monarch. On the one hand, the Commons were surrounded by fantastically decorated prestigious space in honour of their new position; on the other, they were still confined within a royal Palace, under the thumb of the King. The Chamber was used for meetings of the Commons from at least 1343, and then regularly to 1373, when the Chapter House of the Abbey became their preferred location. Within ten years the Commons had shifted to the Abbey refectory and stayed there through to the time when Edward VI dissolved colleges of secular canons and offered the now redundant space of St Stephen's to the MPs. Thus, when the Commons sat in the Painted Chamber in the fourteenth century, it offered far more spacious accommodation than St Stephen's did at any time subsequently.[34]

Connected to the Painted Chamber was Soane's three-bay Royal Gallery, the main part of his ceremonial route for use at State Openings, built in 1823–4. This was adorned with pink columns of fake polished marble known as *scagliola*, and an elaborately tasteful cupola. The end nearest the Painted Chamber was badly scorched by the fire, but most of the neoclassical Gallery and the *Scala Regia*, or processional staircase, survived; its escape was due to the thick stone structure of the Painted Chamber and the strong party walls Soane had installed around his new wing, including the Lords Library. His new buildings beyond, in Cotton Garden, built in 1826 and 1827 also survived: the Lords committee rooms, and the administrative quarter known as the Parliament Office, 'where there are many papers, are chiefly preserved'.[35]

At midnight, John Rickman had reported to his wife that he thought Westminster Hall was very nearly secured.[36] By 1.45 a.m. Westminster Hall was looking safe; the nearby flames having been subdued in the House of Commons.[37] At two o'clock, 'by the blessing of the Almighty', Mrs Rickman was able to feel confident the family house had survived: 'the high tide gave power to the floating engines [*sic*] to throw volumes of water upon the blazing buildings: at the same time came a N. wind, which turned the flames and saved the Hall!—and us—till this from a lack of sufficient water, all must have gone'.[38] Finally, at half past two, the spectators outside began to realize for themselves that Westminster Hall was, at last, safe. The fire still burnt furiously among the ruins of the other Commons and Lords buildings, but its power to do further mischief had ceased, and the flames were now confined within the walls of buildings already destroyed. By three in the morning the fire near Westminster Hall was wholly subdued. It still raged violently in the area of the King's Entrance at the far end of the Lords frontage,[39]

but William Hart, an official in the Comptroller's Office in the Exchequer, considered that 'the burning of the Houses of Lords and Commons was an awful sight at our office being so near but as Westminster
Hall was saved, which was worth a hundred of the other buildings,
I must say I felt quite pleased at the narrow escape it had with the fire
burning quite round the south end of it'.[40]

As well as the objects which survive from the Painted Chamber,
another relic—this time actually saved from the fire—was an oak
table alleged to have been stained with the blood of Spencer Perceval
(1762–1812), the only British Prime Minister ever to have been assassinated.[41] The table was therefore doubly significant. Perceval was
fatally shot in the Lobby of the House of Commons by a disgruntled
merchant called John Bellingham on 11 May 1812. Bellingham har-
boured a grudge against the government for abandoning him when
arrested for debt in the Russian port of Archangel some years before.
After spending seventy-two days in a deeply unpleasant, rat-infested
gaol, he was then placed under house arrest at the English College of
Commerce for two more years. The British consul and various MPs
had failed to take up his cause; coolly shooting the Prime Minister
was, he felt, simply the most effective way of expressing his frustra-
tion, and getting redress of his grievances considered at the highest
level. On being hit in the chest by the bullet Perceval cried, 'I am
murdered!' and fell face forward on the ground. He was then carried
into the Speaker's Secretary's room, off the Lobby, and placed sitting
upright on a table there, supported on either side. He made no sound,
just a 'few convulsive sobs', dying almost immediately. Once he was
pronounced dead, he was removed from the table, and laid on a sofa
in the Speaker's drawing room. The corpse was later transported to
Downing Street at 1 a.m. in the morning.[42] Joseph Hume—who seems
to have got everywhere—was present at the time of the murder, and
played a major part in the immediate aftermath of the killing. He ini-
tialled documents found on Bellingham's person to confirm their au-
thenticity and gave evidence to the investigative inquiries afterwards.[43]

The oak writing table in question was therefore 'greatly prized', and excited 'the greatest curiosity' to view it in the days after its removal from the burning building. It was said to be stained with a sixpence-sized drop of the Prime Minister's blood. 'Great value has always been placed on this relic, and the table has ever since been kept in the room appropriated to the Speaker's Secretary; at the present it is one of the avenues, and appears to be a great object of curiosity', said *The Times*, suggesting it had come to rest in an outdoor location.[44] And yet by the time an inventory was taken by the Office of Woods a few days after the fire a 'large wainscot table (Percival) and green cloth cover' was to be found inside the Palace, on the staircase leading to the Speaker's eastern state apartments.[45] Meanwhile, the Table of the House, reputedly designed by Wren, was also said to have been saved. This had been used from the eighteenth century onwards, and was once the place where the Clerks of the Commons sat when in the Chamber, in front of the Speaker's Chair.[46] This is likely to be the huge square mahogany table which today sits in one of the reception rooms of today's Speaker's House. How this survived is unclear—it is an enormous piece of furniture, which, even if disassembled in time, would have taken a lot of men to carry it out. But it was not in fact in the Commons Chamber by 1834. It may actually have been on the ground floor of the chapter house of St Stephen's, the 'lower oratory', which survived the fire. There is mention of a 'large mah[ogan]y table with 12 drawers in (Oliver Cromwell's)' in the inventory of the furniture stored there after the fire.[47] The two tables, oak and mahogany, seem to have become conflated with one another in the years following the fire. It is not clear where the oak table from the Lobby now is, and popular myth today is that the mahogany Speaker's House table is the one allegedly stained with blood.[48]

Before 3 o'clock had chimed in the nearby streets, there was scarcely a person to be seen in the area except for soldiers and firemen. The crowds which had for hours packed the surrounding alleys and rooftops had quietly dispersed: up to the last, there had been no disturbance, and now the only sound heard was the crackling of red hot timbers, and the heaving of the fire pumps.[49] Edward Villiers had left the Rickmans' house. 'Of course,' he wrote to his mother, a few days later,

> the fire is the engrossing topic; the accounts in all the newspapers are so very full and correct that there is no use in repeating them. I got also a most splendid view of the fire which was burning all around the house. Had I not seen half Constantinople burnt down I would say it was the finest sight I had ever seen, and here also there were peculiar beauties which the other could not have, such as the lighting up of the Abbey, a more beautiful sight than that never was beheld. All the attempts to arrest the fire were for hours unsuccessful; they deserved to be, for they were really contemptible considering the age in which we live, nothing ready, nothing effective when it was ready, and no management whatever. Nothing of great value is lost, and nothing which cannot be re-placed—so as the glorious old Hall is saved (and it really was almost a miracle that it was), I don't so much mind, and nothing is known as to its origin, but the evidence which they have had at the Home Office is all in favour of accident, some stoppage in the flues. It certainly how-ever burst forth in three places at once. The people gave three cheers when the roof of the House of Lords fell in. The King has, I believe, offered Buckingham Palace. This is a true and particular account of all I know on the matter. It is still burning but quite subdued, and they are emptying the Thames upon it.[50]

In the Rickman household all was finally quiet. Mrs Rickman retired to bed at three, exhausted (her husband having 'pronounced victory on the flames'), and slept soundly until eight the following morning.[51] Frances Rickman sat down at a table, and commenced writing letters to relatives to send by the early coaches, since the evening coaches had spread the alarm far and wide across the country. Her 'pacquets' suc-ceeded very well, flying ahead of the bad news speeding towards the provincial papers.[52] Delayed shock only came later when, on the fol-lowing two nights, she was so anxious about the persistent bursts of flame that kept rearing up and threatening Westminster Hall again that she slept with her bedroom window shutters and door open to enable a rapid escape, Cluny (presumably her favourite pet) by her side, and with her valuables ready-packed in a basket for a quick getaway. She was thankful there was no one ill in the house whom it would have been difficult to evacuate at short notice. The house and the Rickmans' possessions stank of fire for days afterwards. She began her letters first with one to her sister Anne, who was visiting their uncle in Chichester, later asking her to return it so she could send it on to her aunt, 'for I cannot say it all again…how much more I could like to say'.[53]

Friday, 17 October 1834   3.30 a.m.

Thank God we seem all safe

PALACE YARD 17TH OCTOBER

½ past 3 a.m.

Thank God, my dearest Anne, after near eight hours dreadful doubt, we seem all safe, though I am still partly lighted by the still blazing House of Commons! I fear you will hear of the awful fire before this reaches you. I have been much alarmed for us, indeed, and you will imagine it has been the most awful thing it is possible to conceive. I will give you as collected an account as I can, for my legs ache and I could not sleep, so I may as well write. After dinner, at ½ past six this evening, Papa and Mamma taking a nap, in came Ellis—'Think, Miss, there's a small fire broke out at the House of Lords.' I said 'Come with me to the leads to see it,' and there, even then, a volume of flame was blowing towards the Wildes'. Papa at first thought it could be got under, but soon it fearfully grew, and we had little doubt that the Hall would catch. The House of Lords we could not see, but soon heard that it and Mr Ley's and the Library were destroyed: then the flames burst from [the] House of Commons windows, and sooner than I could believe the interior of that was destroyed.—Now see my view, the west window in bow room my prospect, front state rooms of Speaker's remain entire (outwardly), red smoke rises from the quadrangle, and the open House of Commons arches (ruined like Fountains Abbey) are filled with an orange light, nearly the whole of [the] south end of Speaker's is destroyed. Poor Papa will grieve for the Library; Mr May is lately returned and Mr Vardon came up last night! But—for the woeful effects on us—I first ran to the W[ilde]s' who with Mr Gurtkin were in agony that, as first appeared probable, they would be burnt; even then blazing papers were floating over and in their garden. I brought some valuables to our house—but soon the tide turned and we were in danger, so Papa thought we should put things together and some went to Mr Mundell's with the Wildes' goods and Mr Bitts'—Poor Mamma was much overcome at first, but that made me stronger, as I felt I must look to everything, Papa being then rather provokingly easy; by this time we had many helps and constant knocking at the door. I chained it so as to prevent any dangerous visitors—the Yard was full and the River mud (for tide was low) crowded with people—presently in came poor Mr Manning who had spent the day out—we had a most affectionate kissing meeting—he saw it in Oxford St and rushed down. Ellis, Mr Pritt, Apps, James (Telford) the Dean of Ripon's servant sent to help. Mrs Doctor Holland's coachman and footman here, when came a knock, and Henry Taylor answered my 'your name, if you please', before I let him in. He had a tall elegant friend with him, Mr Edward Villiers, and they insisted on being active chief managers under me, and worked furiously, H.T. getting coaches, taking their number, filling them, and sending a servant on the box of each to unload, to carry goods to Mr Seaton's, and on Terrace Mr Bacon's—Ad. Lee's and Mr Wilde's, for the books were tied in sheets, drawers emptied, everything dismantled. Here (bow room) only a few chairs, sofas and the table remain, your drawers I sent down not having the key, now the Wildes were not in danger and I carried all dear W[illiam]'s things there—Thomas soon packed the plate—Fancy the whole house dismantled, H.T. and friend working away; I shall never cease to respect his judicious management and energy. Mr Payne came, beds offered us on all sides, Mr Seaton in and out—Mr Robertson, Walker, Hansard, Roland, Tatershall,—as often as Papa and I could we went to comfort the W's; Captain Colquhoun was directing on the Speaker's House. They knocked in the roof. The furniture was all thrown out of the windows, even china, mirrors. Mr Fairlie left yesterday! The Speaker's son was just dressed and going out to dinner—Mr Roland who was there said he was earnest about the plate and rather indifferent to all the rest—the pictures (Speaker's) I believe are moved but being no good

209

superintendant there, I fancy the devastation is dreadful. It is feared one poor maid is burnt, she is missing, but I think probably ran away; one poor man was carried off both legs broken; a soldier suffocated, a fireman killed so we hear: the police order was beautiful, the Horse Guards down, and H.T. as he came met Lord Munster, and considerately asked for a dozen soldiers to stand at our door. What a subject for his next poem!

I am truly thankful that I was enabled to use more energy than I can now believe possible. Truly strength is given in the day of trial. Poor Hannah was as white as a sheet and Jane very frightened. Dear Mamma soon became cool and packed in the trunks as if going on a journey. Mr Manning established himself in two chairs in [the] long passage. Papa and Mr Payne took me out to the corner of Palace Yard to see the Abbey, such a grand sight as I may pray I may never see again; the bright moon in dark clouds, and the clear red and blue and yellow light. Oh! no one who did not see it can picture it. I rejoice my dearest Anne you were spared this scene we are all very glad you are away— Oh! When will the house be right again—but no complaints—what a mercy that it happened before night—and no lives from surprise can be lost—I truly rejoice in the Wildes' safety—do you feel [sic] that all our things that are not gone are in the Hall. The featherbeds tied up ready to go—so you may fancy our real danger. They sent off to Speaker at Brighton, he knew it by 12 at night—a great mortification to them and Mr Ley—it would have been singular if thus the old Exchequer had ended! As I walked by the Hall its S[outh] window was a blaze of light—now I think it hardly safe, still there is some fire near it very vigorous. What an event in the nation! And Ellis near the first to discover it—it is said to originate in a Gentleman writing in an upper House of Lords room. Hannah is in your bed—Mama in hers, Papa asleep on this sofa. Mr M. [is] in the little room next [door] and comes into me whenever there is a burst of noise, he had brought me a little shrub* that I may express myself con brio, *his most affectionate regards to you, he could not help making a few good puns. You will be astonished that H. Taylor should be the hero. I should think Speaker will be up soon. I hope the Gobelin tapestry is saved. Fancy the Spanish Armada and all etc. destroyed! Mr Apps who stayed last (till 3½) the others all till past two—fears much Mr Vardon had some bonds in the Library table drawer. The Whigs and Reform Parliament will indeed be remembered.*

---

* A drink made with fruit juice, sugar, and a shot of rum or brandy.

We need not look for a new house in the neighbourhood. God bless you and Uncle, I cannot say more—Ever your F.R.

½ past six

Daylight, and after a hard fight to save the Hall, the fire is all out. Papa thinks saving the Hall next a miracle. Speaker's Garden filled with furniture. I write by Coach to Aunt Anne. The 12 hours do not seem two.[1]

14

Guy Faux has rose again

I N THE DARKEST hours before the dawn, the Palace presented a shattered and eerie spectacle, still luridly lit by the flames and an immense mass of glowing timbers on the ground. As it cooled, the medieval fabric of the original buildings emerged through the smoke: a weird parody of its former self, stripped of its later accretions by Wren, Wyatt, Smirke, and Soane. The intense action of the fire had covered the bricks, tiles, and slates 'with a variety of vitrifications, and ashes of various hues, assuming in one case the appearance of perfect glass, in another of polished ware, and in a third of vegetable moss, whilst the liquid lead hung in some places like enormous icicles, or masses of watery weeds'.[1]

A further discovery added to the strangeness of this topsy-turvy world. Lying among the ruins were the red-hot weights and measures of the kingdom. These were the national master standards against which all local copies in the shires were calibrated. Originally ordained by Magna Carta in 1215, the standard yard was used to compute distances, volumes of land, and lengths of cloth: feet and inches; poles, perches, and rods; furlongs, miles, and acres. The standard pound weights allowed the calculation of ounces, grains, and drachms; of liquids (gallons, quarts, pints); and of heaped measures (bushels, pecks, quarters, and sacks). They were a check on the honesty of tradesmen in all manner of commodities including medicines, grain, beer, spirits, salt, oil, fruit, vegetables, feed, coal, dry goods, fish, lime, bullion, and even the manufacture of weights and scales themselves.[2] The standards had been housed at Westminster in the custody of the Clerk of the House of Commons, although their verification fell to officials of the Exchequer.[3] Over the summer of 1834, the Commons had been considering amendments to the law, and the eighteenth-century standards may not have been in their usual location at the time of the fire. Charles Rowland, a Journal Office clerk, found them in the rubble and made a list of what he discovered. The 1753 and 1760 brass standard yards, and their yard-beds, were all discoloured by the heat, and their gold studs— between which the measure was taken—were melted out; the 1753 yard was additionally bent out of shape. On inspection of the 1760 measure, it was found 'impossible to ascertain from it, with the most modest accuracy, the statutable length of one yard'. The 2lb, 4lb, 8lb, 16lb, and 32lb weights cast in 1758 were all discoloured and compromised. The legal standard of one troy pound was missing.[4] There was no longer any means for the nation to confirm definitively how heavy an object was, or the length of any distance.

It was 'almost incredible that so much destruction could have occurred in so short a time', wrote *The Examiner*.[5] The printers at *The Times* were busy at their presses, producing the first reports of the devastation the fire had left in its wake. Later that same morning, readers at breakfast tables, in coaching inns, and in fashionable clubs absorbed the news that the House of Commons, House of Lords and the Painted Chamber, the Commons Library and Mr Ley's House were all burnt out, with only their bare walls still standing. The front of the Commons Library had fallen in; and half of the Speaker's House, including the cloisters, was burnt. The Parliament Office (at the western end of the

House of Lords, entered by the Abingdon Street gateway opposite the Star and Garter, readers were reminded) was saved, together with all the books and papers it contained, and all the books from the Lords Library. They were able to read that some of the Commons' books and the Speaker's furniture had been removed early by the police, and placed in the yard and terraced garden adjoining, covered over with carpets and tarpaulins. A £900 marble mantelpiece at the Speaker's had been taken down and removed to a place of greater safety. They discovered that Soane's King's Entrance in Abingdon Street and the *Scala Regia* were preserved, communication with the rest of the burning building having been cut off. Westminster Hall, 'for which the greatest anxiety was evinced by everyone', was safe, announced *The Times* triumphantly, although everything beyond its great south window appeared to be a complete ruin. The glass of the great south window was of course broken, but the mullions remained intact. The Courts of Law were uninjured, or believed to have sustained only some very trifling damage.[6]

James Braidwood mused on the causes that had made the Palace fire proceed so rapidly. He had no doubts that they were:

1. The total want of Party Walls.
2. The Passages which intersected the Buildings in every direction and acted as funnels to convey the fire.
3. The repeated alterations in the Buildings which had been made with more regard to expedition than security.
4. The immense quantity of timber used in the interior.
5. The great depth and extent of the Buildings.
6. A smart breeze of wind.
7. An indifferent supply of water which although amply sufficient for any ordinary occasion, was inadequate for such an immense conflagration.
8. My own and the firemen's total ignorance of the Localities of the place—in private dwellings, Warehouses, or Manufactories, some idea may generally be formed of the division of the inside of the premises from observing the appearance of the outside, but in the present case this rule was useless.[7]

Braidwood considered that the origin of all fires could be refined down to just five causes: inattention in the use of fires and lights; improper construction of buildings; use of furnaces or close fires for heating or for mechanical purposes; spontaneous ignition; and what he termed incendiarism. Nineteen out of twenty cases in his experience were simply 'the result of remissness or inattention'.[8] The danger from furnaces or close fires (that is, solid fuel stoves), whether for heating, cooking, or manufacturing purposes, was very great, he believed. No flue should ever be permitted to be so used, he warned in his publications, unless it was specially built for the purpose. This was because, in a flue, all the air being drawn through it had to pass through the flames first. The draught thus became so heated that, unless the flue was properly built, it was 'dangerous throughout its whole course...in open fireplaces the quantity of cold air carried up with the draught keeps the flue at a moderate heat, from the fire upwards...this is the safest possible mode of heating...heating by hot air, steam and hot water are objectionable [because] the flue used is generally that built for an open fire only; and second, the pipes are carried in every direction, to be as much out of sight as possible'.[9]

Braidwood also understood the psychology of crowds. Most people's experience of a large fire would be as spectators. This left them wholly unprepared to react sensibly if a fire were ever to ignite in their homes or workplaces. As a disinterested observer a person might, after the initial fright and confusion on a fire's appearance, consider that a conflagration 'may from its novelty or grandeur, if the fire is extensive, be still worth looking at for a little, but much of the excitement is banished with the confusion; and if the fire and firemen seem to be well matched, the chief interest which is excited in the spectators is to ascertain which of the parties is likely to be victorious. Few people, comparatively, have thus an opportunity of witnessing the terror and distraction occasioned by the first alarm of fire, and this may probably account for the apathy and indifference with which people who have not seen this regard it.' Yet if a fire were ever to affect them and their loved ones directly such people would be completely unprepared: 'coming on them like a thunderbolt, they are lost in perplexity and terror'.[10]

George Rose, clerk to the LFEE, found he had quite a lot of difficulty recording such an unusual fire sensibly in his daily ledger of incidents. Struggling to fill in all the information later that day, he wrote:

> Hour: 6½ PM
> Place: Parliament St
> Name of Occupier of Premises: Peers of the Realm
> Business: The House of Lords & Commons
> Name and Residence of Landlord: Government
> Supposed Cause of Fire: Not Known
> Where insured: Not Known
> No. of Policy: ——
> Water, with Name of Company: Chelsey
> Damage: Nearly the Whole of the House of Lords and Commons and
>   Part of the Contents Therein
> Called by a Stranger
> Extinguished by the Firemen and Others
> Gas [on the premises]
> Further Partickulers Will be Made this Day.[11]

But that was all still to come. In the hours before dawn, fresh engines and additional supplies of men were still arriving at the scene of devastation, and a continual volley of water was still in action. 'More vigorous exertion and more active zeal were never witnessed,' said *The Times*,

> but it must be confessed that our ordinary engines are totally incapable of contending with such a conflagration as that of last night, and that our fire-engine system wants the great element of efficiency—a general superintendent: each fire office acts according to its own view: there is no obedience to one chief, and consequently where the completest co-operation is necessary, all is confusion or contradiction. We impute no blame to the fire-offices or their men: the conduct of individuals was above all praise, but the want of a general leader and director must have been in the course of the evening as evident to them as it was to the discerning portion of the spectators.[12]

James Braidwood remained on site until five o'clock. At that hour, the military and police were relieved, the new shifts taking over now the crisis had passed.[13] The engines returning home would need to be serviced the day after—oiling, cleaning, and the repairing of any damage occurring as soon as possible—so that they would be absolutely ready to be called out again when needed next.[14] The British Fire Office sustained wear, tear, and damage to its engine to the tune of £21 7s. and billed £30 15s. for use of its private firemen.[15] The County Fire Office charged the government nearly £13 for repair of its engines and its firemen cost the Woods and Forests £6 11s.[16] Damage to the St John's, Smith Square parish engine came to £5 12s.[17]

He and his men had done their best, and to Braidwood must go much of the credit for saving Westminster Hall, and the Lords Library and Parliament Office (the administrative quarter of the House of Lords). In the days that followed, however, there was much criticism of the overall management of the firefighting. Joseph Hume's comments were typical: 'In the zeal, order and numbers of the military and police, and the daring of the firemen, nothing was wanting; but in the power of the engines, the quantity of water, and the general superintendence of the whole operations, a most lamentable deficiency was apparent.'[18] The fire was enormous, complex, and terrifying. The resources at Braidwood's command were limited. Today a fire of that size would certainly involve many more firefighters than he employed, even with the advantages of all the modern equipment, technology, and chemical suppressants available. In addition, he had no control over the actions of the private insurance companies outside the LFEE, and was unable to prevent the interference of senior politicians and royalty in the operations. Stung into defending his own conduct, Braidwood wrote to *The Times*:

> I acted then as I have done for the last ten years viz., heard all the advice given me as patiently as I could, and acted on such as seemed to me better than the ideas I had first formed on the subject, visiting the different positions, to see whether the increase or diminution of the

fire required them to [be] altered, attending to the supplies of water, refreshments to the men etc...I have every reason to be satisfied with the zeal and alacrity displayed by the Foremen, Engineers and Firemen belonging to the [London Fire Engine] Establishment. Other fire engines attending, from insurance offices, parishes or private individuals were each acting under the orders of their respective foremen.[19]

'My Dear Tom,' wrote onlooker Charles Fremantle to his brother a few hours after leaving the fire,

we returned to Town last night...in time to witness the awful fire which took place, which has destroyed entirely both houses of Parliament & part of the Speakers house. I was dining...when the servant told us & we immediately started off to the spot, where we found the fire had been raging for two hours. A more splendid sight I never witnessed. To stop it was impossible as the wind was blowing very strong from the S.W. & the Engines were of little use. Most of the Gentlemen exerted themselves to save the public papers & documents &...the library of the House of Commons & half of that of the House of Lords, but as many of the papers were thrown out of the windows in the street they fell into water & dirt & are much defaced...The general idea is that it was [an] accident, but some people declare that as the fire broke out at both ends at once that it must be the work of an incendiary & there is also another story of a quantity of matches having been found...All the Speaker's furniture, books, etc, have I fear suffered very much, as everything was thrown out of the windows & taken out of his house as fast as possible...I was at the fire most of the night & worked very hard at one time, & today I have been all over the ruins & a more frightful mess you cannot conceive, it is smoking & parts of it still burning...It is most fortunate that...the question now as to the necessity of new houses of Parliament is set at rest.[20]

The local hostelries had been drunk dry. Apart from quenching the thirst of the crowds, there had been the rescuers to supply, both the volunteers and soldiers pumping the engines and the firemen needing refreshment. Beer was obtained from the Star and Garter in Old Palace Yard, the Horseshoe and Magpie in Bridge Street, and the Westminster Hotel. The owners were compensated accordingly, with the drinks

PLATE 17   A thrillingly naïve, but nevertheless highly accurate view of the fire from the Lambeth bank of the river, the flames and sparks blown briskly from the south-west, showing the crowds on the muddy shore, the full moon, and the lit-up towers of the Abbey behind.

PLATE 18   John Rickman (1771–1840), Clerk Assistant of the House of Commons.

PLATE 19   Charles Manners-Sutton (1780–1845), Speaker of the House of Commons.

PLATE 20   View of Old Palace Yard facing east, probably towards midnight. In the foreground, huge piles of salvaged furniture and other items are covered in tarpaulins.

PLATE 21   The interior of Westminster Hall at the height of the danger. Scaffolding and
building materials associated with Smirke's repairs remain in place, while rescuers scale the
walls on the eastern side adjacent to the cloister and Speaker's House. At the far end, firefighters positioned by the great south window attempt to prevent the flames from attacking
the famous hammer-beam roof, powered by a water supply from the two attached LFEE fire
engines. In the foreground on the right, a thirsty worker takes a swig of much-needed beer.

PLATE 22   A sketch, reproduced in a number of newspapers, showing the Floating Engine in position about 1 a.m., in front of St Stephen's, emblazoned with the name of its former owner, the Sun Fire Company.

PLATE 23   Looking south over the cloisters, towards the ruined St Stephen's Chapel, with the Speaker's House on the left-hand side. Probably drawn after daybreak on 17 October, as firefighters continued to damp down fires breaking out on the roofs..

PLATE 24 Old Palace Yard the morning after the fire, before the hoarding went up, showing workmen carting away ruins and spectators promenading. Wyatt's frontage of the Lords
has fallen in, revealing the arches of the Lesser Hall. Firemen, with a ladder, are still tackling
outbreaks of flame in the Commons, where the roofs have collapsed and all the windows
have smashed. This rather sanitized sketch was published on 20 October 1834.

PLATE 25 Close-up view of the collapsed west front of the House of Lords revealing the
walls of the Lesser Hall, the former Lords Chamber, in a watercolour by the prolific Robert
Billings.

PLATE 26   The ruined river frontage the day after, showing from left to right: the undamaged Parliament Offices of the House of Lords by Soane; the ruined Painted Chamber; the collapsed bays of the House of Commons Library; Mr Ley's House, gutted; St Stephen's; and, behind the trees, the Speaker's House.

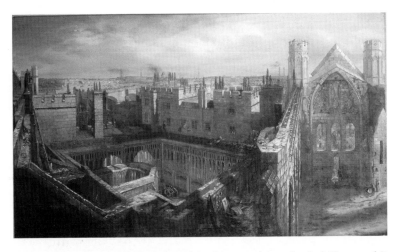

PLATE 27   George Scharf senior's view of the chimneys, cloister, Speaker's House, and St Stephen's Chapel, drawn over a period of three weeks beginning with the morning after the fire. This painting was part of a diptych whose other half (with the southern panorama round to the House of Lords) is missing. Note the abandoned fire hose in the ruined Chamber and the fire engine still damping down the cloisters.

PLATE 28   The Painted Chamber in use as the Court of Requests before the fire, looking east. The original wooden paterae are faintly visible on the ceiling, and some original
medieval paintings can been seen in the window splays on the right; the remainder are hid-
den underneath the whitewashed walls.

PLATE 29   The same view of the Painted
Chamber looking east after the fire, including an artist sketching the ruins. The
walls were in fact still stable enough to
allow the building to be fitted up later as
the temporary House of Lords.

PLATE 30   Archival hero: Henry Stone
Smith (1795–1881), committee clerk in the
House of Lords, whose swift and decisive
actions on the night helped to save many of
the historic records of the House of Lords
inside the Palace.

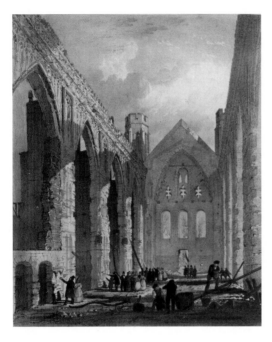

PLATE 31 Sightseeing the ruins of the old House of Commons, where, the roof and floors having fallen in, the old shell of St Stephen's Chapel is now revealed in all its glory. The line of the floor and ceiling of Wren's refurbished House of Commons Chamber can clearly be seen on the east wall, and his bricked-up east window is still intact.

PLATE 32 Clearance and demolition operations on the west front of the House of Lords in November 1834. By this time the unstable walls and towers in front of the old House of Lords Chamber have been pulled down and the Lesser Hall is fully revealed with its semicircular 'Dio-cletian' windows, ready to be used as the temporary House of Commons.

PLATE 33 Frances Rickman's letter to her sister Anne, written at 3.30 a.m., with a sketch of the fire in St Stephen's Chapel as seen from an upper floor of their house.

PLATE 34   The shored-up Painted Chamber and the shell of St Stephen's Chapel in 1835 by Robert Billings, looking northwards along the river side of the Palace. The Commons' Library and Mr Ley's house in between have been demolished and cleared, and St Stephen's is shortly to follow.

PLATE 35   Barry and Pugin's new Houses of Parliament rising up around the old, as seen from a balloon in the late 1840s. The scale of the new Palace has been exaggerated, and Westminster Hall and the Law Courts have disappeared altogether, but the Lesser Hall can clearly be seen, as can the King's Entrance by Soane.

bill totalling £11 10s. 2d. at a time when a quart of beer—that is, two pints—cost a penny.[21] Assuming the publicans sold the beer to the authorities (maybe by the barrel) at retail prices, rather than wholesale, then that equates to 5,524 pints of refreshment during the fire. Mr William Nichols, of the Westminster Hotel, had supplied additional coach hire to the tune of £1, and more beer to £3 10s. Furthermore, Jane Coney, landlady of the Red Lion public house in Westminster Street, was paid another £18 16s. for beer (some 9,216 pints) supplied 'by order of the Speaker'.[22] One explanation for this retrospective payment might be that this was the drinks bill for those who worked at Speaker's House on the night. Or the beer could have been for the workmen who were associated with the return of furniture, pictures, books, and records—or even salvage operations—in the days following the fire. Whatever it was for, it was clearly very thirsty work.

As the flames began to die down, the outlines of the devastation continued to emerge through the smoke. The remains of the Painted Chamber and the Commons Library presented 'such a mass of ruin that it is difficult to trace the site on which they stood'. The frontage towards Abingdon Street with the exception of the King's Entrance and the Jewel Tower on the opposite side was a 'confused heap of ruins'. The arcade in front of the House of Lords had disappeared and, the greater part of the façade constructed by Wyatt having collapsed, the ancient medieval walls of the Lesser Hall—the southern gable end of the Chamber—were now visible. A stack of chimneys at its northern extremity stood alone, apparently in a tottering and dangerous position. The walls of the wing occupied by the Commons, formerly Bellamy's, several of the committee rooms, the gallery, and, on the ground floor, the Members' Entrance and waiting room were all that remained of the Stone Building, the rest was burnt out, with windows gaping and the roof caved in. Its northern wall formed the boundary of fire damage in that direction, the Rolls Court which adjoined it being untouched and the other Law Courts 'uninjured to any considerable extent'. Towards the river, the

219

Parliament Office was unaffected, 'beyond such damage as the hurried removal the furniture, books and papers must have occasioned'. The Painted Chamber, the Library of the House of Commons, Mr Ley's House, and the House of Commons Chamber were all 'gutted of every particle of the timber, a smoldering mass at the bottom presenting the only remains except the bare walls'. Three or four of the rooms of the Speaker's House 'are also consumed, as well as the State Dining Room, which is of course demolished, as it was under the House of Commons. The fire was fortunately checked in that direction, or the destruction of Westminster Hall would have been inevitable.'[23]

The official report from the Office of Woods and Forests, issued by the end of that day, laid out the damage in starker but more definitively accurate terms:

House of Peers: The House, Robing-room, Committee-room in the west front, and the rooms of the resident officers, as far as the Octagon Tower at the south end of the building—totally destroyed. The Painted Chamber—totally destroyed. The north end of the Royal Gallery, abutting on the Painted Chamber, destroyed from the door leading into the Painted Chamber, as far as the first compartment of columns. The Library and the adjoining rooms which are now undergoing alterations as well as the Parliament office and the offices of the Lord Great Chamberlain, together with the committee rooms, housekeeper's apartments, &c in this part of the building, are saved.

House of Commons: The House, Libraries, committee rooms, housekeeper's apartments &c are totally destroyed (excepting the committee-rooms Nos 11, 12, 13 and 14, which are capable of being repaired). The official residence of Mr Ley (Clerk of the House)—this building is totally destroyed. The official residence of the Speaker—the state dining room under the House of Commons is much damaged but is capable of restoration. All the rooms from the oriel window to the south side of the House of Commons are destroyed. The *Levée* rooms and other parts of the building, together with the public galleries, and part of the cloisters, very much damaged.

The Courts of Law: These buildings will require some restoration.

Westminster Hall: No damage has been done to this building.

Furniture: The furniture, fixtures and fittings to both of the Houses of Lords and Commons, with the committee rooms belonging therein, is with few exceptions destroyed. The public furniture at the Speaker's is in great part destroyed.

The Courts of Law: The furniture generally of these buildings has sustained considerable damage.[24]

Despite this gloomy report, fire did not raze most of the old Palace to the ground. Charles Barry did that. The medieval walls still stood in most cases. An engraving from the late 1840s, illustrating Westminster and London from a balloon floating somewhere over Pimlico, shows the half-built new Palace springing up around the tiny old one. While this view deliberately over-emphasizes the enormousness of the new Palace (and entirely obliterates Westminster Hall), it does indicate that a substantial part of the old Palace did survive in battered shape through to the early 1850s.[25] The shells of the 'totally destroyed' Painted Chamber and the House of Lords were still sound enough, a full survey indicated, to be reused. But there was no going back to how things were before. It was acknowledged that these refittings could only be temporary measures. The House of Commons was beyond repair, but even there most of the walls still stood, when shored up. The 'improved' St Stephen's Chapel, as John Rickman thought of it, initially made an excellent ruin: 'the crypt and beautiful cloister adjoining prove the efficacy of arched roofs, as they are imagined, even to the colouring of the keystones and bosses, so you must not blame me for vilifying the wooden substitute (a kind of architectural fraud) at York, which caught fire in 1829, and has cost £100,000 in reparation'.[26]

The family's friends expressed astonishment at the beauty of the walls and the Speaker's Cloisters as they toured round some days later.[27] Drawings of the ruins at St Stephen's after the fire indicated that after Wren's wainscoting and suspended ceiling had been burnt away, Edward III's chapel had revealed itself in all its glory. The soaring stonework, elegant tracery, and richly decorated capitals delighted sightseers, and, amazingly, it was discovered that even James Wyatt had not destroyed all the wall paintings in 1800 as mural decorations were uncovered by the fire in the upper reaches of the walls: fleur de lys, stars in lozenges, heraldic lions, and figures of saints and biblical characters were clearly visible.[28] St Stephen's was pulled down in the late summer of 1835, after Barry expressed alarm about its instability.[29] Just a decade or two later, contemporary fashions would have changed and it might well have been preserved.

The names of some of the ordinary people who had helped with the rescue were soon collated and compensation offered to them by the Office of Woods and Forests. Police Constable no. C120, William Sullivan, fractured his right hand. He was paid £5 compensation following treatment in Charing Cross Hospital.[30] John Slater, who was 'much hurt and obliged to attend the Westminster Hospital for several days', received £5 18s. 6d. for his lacerations.[31] Thomas Nesbet, a tailor of Berwick Street, Soho, received £1 1s. because a contusion on his right hand meant that he was unable to work for a fortnight afterwards.[32] Philip Ellis, a waterman, sustained an injury in removing a large clock, probably from the Speaker's House. That cost the government ten shillings.[33] John Kennedy, a helpful hackney-coachman, had an accident too, leading to a loss of earnings of £4.[34] Ralph Raphael, who had been taken to Westminster Hospital with a suspected fractured skull, was compensated for a broken leg and other injuries to the tune of £20.[35]

Four carters were paid £10 6s. 8d. by Henry Stone Smith for transporting books. Perhaps they went to Mr James Bigg, bookseller of Parliament Street and Mr William Isacke, bookbinder of Carter Street, who also received payments for assistance. May & Phillips, carpenters of Smith Street, were paid nearly £54 for the work of their men in cutting away the Speaker's roof, while Robert Johnstone, the master mason whose thirty men had intervened so successfully in Westminster Hall, billed the Office £59 for their work. Fourteen carpenters under Mr Griffiths and thirteen workmen under Mr Cowdray—presumably government contractors—received ten shillings each for their help. At least twenty-one other labourers or watermen from across Westminster were paid between ten shillings or a pound each: quite a haul. (As a point of comparison, £1 of earnings in 1834 would be worth almost £750 today.[36]) Another four were paid on the recommendation of the Court of Chancery office keeper—the lawcourt closest to the blaze in Westminster Hall. Robson & Estall, a plastering company, were able to replace ladders worth £7 19s. Ordinary individuals who rolled up their sleeves to help had lost or ruined watches, hats, coats, waistcoats, silk handkerchiefs, including Mr Fache and his son, respectively confidential messenger and messenger to the House of Lords. People from all walks of life had helped: politicians, aristocrats, butchers, lawyers, tailors, bricklayers, and military among them. In some cases, usually gentlemen who could afford it, they refused to be refunded their losses—such as Mr Pearce, a surgeon of Marsham Street and Mr Robert Munden of Great George Street.[37] Many hundreds more remain unknown and unnamed.

Then claims for compensation from Parliamentary staff began pouring into the Office of Woods and Forests. From the Lords, Black Rod—Sir Augustus Clifford—asked for 'between £80 and £100' to replace his dress clothes and books. Mr and Mrs Mullencamp claimed £218 18s. The doorkeeper and his wife (the housemaid who had first discovered the fire) were considered to be 'among the principal sufferers'. The entirety of the Mullencamps' property was destroyed, items both of instrinsic worth and of sentimental value. They were not insured. Mrs Wright submitted a request for £135 10s. to replace her furniture and other goods.[38]

In the Commons, Mr Ley asked for £45 for his servants' furniture.[39] Mr Bellamy claimed for £417 for loss of his own goods, and £136 16s. 6d. on behalf of his six servants—including Nicholas, the refreshment room butler—who lost their uniforms. But Bellamy had been covered by a policy and was only claiming its shortfall. By insuring his goods, declared the government, he had acted on the principle which it conceived to be sound, 'viz that individuals having official apartments have no more right to compensation for loss of fire from the publick than the inhabitants of any other residences have a claim on their landlord for compensation in the event of their residences being consumed by fire'. With regards to the second claim on behalf of his servants (who worked both for the family and for the Members), 'no doubt it is a severe loss to them, but they can have no claim beyond compassion on the publick'.[40]

The claim of over £387 from Mr Leary, Librarian of the House of Lords, was turned down, as was Mr Ley's for furniture, on the 'insurance principle', and for compensation for his servants since they had been privately employed by him. Mr Ley was however given an allowance of £500 a year for three years to fund a residence, or until a new house was provided for him, since that was an emolument connected with his job.[41] Mrs Prescott, Housekeeper to the Parliament Office, had no more claim than the others; Amelia Warren, claiming £24 10s. 6d. for glass etc. destroyed, was informed firmly that she only attended the Lobby of the Commons by permission and had 'no right to have any of her goods under the roof of the Houses of Parliament and has certainly no claim for compensation'.[42]

Mr James Dobson's £46 10s. claim for the loss of his purse and watch would if admitted 'lead to boundless and unjust preference'. Although Butt, the Deputy Serjeant at Arms, and others had spoken highly of this painter's services on the night, the Office of Works responded that while Dobson's functions may have been useful they were not of 'such transcendent merit' as to justify his and every other person using their presence on that occasion to request remuneration for losses of that description. Of others, they believed it was the Deputy Serjeant at Arms' responsibility to offer assistance directly.[43]

An attempt to reward seven firemen with £1 each for their particular services on the recommendation of Mr Butt and Mr Merryweather, including Birch and Hill, who had saved the Mace (£2 2s. for them), was disallowed. The name of John Hambledon, the most seriously injured fireman, who sustained a compound fracture of the thigh, appears in the accounts of the Office of Works, following an application from the LFEE to compensate him. The space in the book indicating payment is empty.[44]

The total compensation offered for loss and injury was about £156.[45] Jane Wright, Deputy Housekeeper of the House of Lords and daughter-in-law of Mrs Wright, was still arguing with the government over a claim in 1837. 'By the destruction of the Houses of Parliament in 1834, your memorialist lost property to the amount of £400, all she possessed,' she submitted.

> Others suffered at the time, but to no one has the visitation been so severe or lasting as to your Petitioner. Entirely dependent on the fees received for showing the House, her income totally ceased at the moment it was most needed... Under these circumstances, your Memorialist humbly implores your merciful consideration of her case, and that you will be pleased to grant her such remuneration as in your wisdom and kindness you may deem meet.[46]

'Mrs Wright the Deputy Housekeeper to the House of Lords appears to us to have no claim whatever to compensation for her losses,' responded the Office of Woods and Forests, sharply.[47]

Poetic contributions about the fire ranged from the elegiac and sentimental to the frankly scatological. One rumbustious verse with a chorus of 'Oh what a flare-up | Crikey Bill, what a jolly flare-up!' tells the story of Bill Jenkins, an out-of-pocket carriage driver, who carries a sense of general disgruntlement with the world:

> The evening advanced, now mind what I say,
> Billy Jenkins was seen to slip slily away,
> And very soon arter, they saw with amaze,
> Both houses of Parliament all in a blaze.
> Down tumbled the busses, the engines they roll'd,
> And the military came with their helmets of gold;

> When a lanky policeman, bawl'd out in a fright,
> We must find out that ere covey what set 'em alight.

The police begin to search, and discover his hat:

> And near it some matches, were strewed on the ground,
> In exact the same spot w[h]ere Guy Fawkes was found.

He is finally discovered stuck headfirst in a privy in the Palace:

> They look'd down the hole, Bill spoke not a word,
> Till they said, Bill, are you dead, no I'm only *inturd*,
> Then they pull him out, and ask'd how he felt,
> Says one, he's as dead as a herring, no, I'm only a smelt.
> I never was in such a strong place afore,
> And I'll take gallows good care, I go there no more;
> You can't think what it is with your different notions,
> To be wrapp'd up in members of parliament motions.[48]

A more sedate and plangent tone was struck by a piece in *The Keepsake*:

> Ruined and gone! How like a feverish dream
> The pictured glare upon that peopled stream!
> The wild confusion, and the startled cry,
> The smoky columns blotting out the sky,
> The pale sick moon, the busy hurrying crowd,
> And the tall distant Abbey, nobly proud—
> All this was real![49]

The tune 'The Good Old Days' was pressed into service for a number of jolly songs by a John Morgan, who produced such literary gems as,

> Fifty five old women swore
> They bet three gallons of gin
> That Old Guy Fawkes was in the flames
> They saw him walking in.[50]

But perhaps the most amusing poem of all is a ballad entitled 'A Conversation between the Abbey & Westminster Hall'[51] which imagines an exchange between the two buildings, almost as if they were two elderly matrons gossiping over a fence in the back yard:

One Friday morn, e'er the moon was gone
Tis true what I'm relating
I stood amazed, at the furious blaze,
Though they were fast abating;
And there, most clear I then did hear
A solemn voice to call
Which began an oration, or conversation,
Between the Abbey and Westminster Hall

Said the Hall, this night, I've had such a fright
The like no man ever knew.
I've had such a roasting, broiling, and toasting
It has put me quite in a stew.
Tho' the rabble did call, in respect for the Hall
But to you, I'll tell the truth,
Tho' my sides they did save, from the furious blaze,
It has scorch'd my ancient roof.

CHORUS
On the Friday morn, when the fire had gone
That had ravaged St Stephen's Hall,
I heard this oration, or conversation,
Between the Abbey & Westminster Hall

Says the Abbey, my friend, I thought you were at an end,
For I heard L—d A——p[a] bawl,
Come lend a hand, the Commons be damned,
But save, O save the Hall!
And there was Middlesex Joe,[b] who not long ago,
For a New House did move I'm told,
Was in a rare pet, and a devilish sweat,
In trying to save the old.

There was Munster Fitz-carnce,[c] had a very near chance,
I'd have given his life for a *furdon*
He swore most stout, that the fire to dout

[a] Lord Althorp, Chancellor of the Exchequer.
[b] Joseph Hume was MP for Middlesex.
[c] George FitzClarence, Earl of Munster.

It would take all the River *Jordon*[d]
But I'm still in a quandary, for they say 'twas an Incendiary,
That made me so precious hot,
What's more, I hear, that Swing is here,
And I shall soon go to pot.

Oh! no, says the Abbey, he'll not be so shabby
As to burn either me or you.
For in ricks or barns and country farms,
He finds plenty of work to do.
But this I hear, though Swing's not here,
There's a man of greater renown;
Fam'd Guy Faux of old, has rose again, we are told,
And the other day came into town.

Says the Hall, I can't see, between you and me
How this fire it first occurred
In the Commons some say, the Members one day
Set fire to the House with hot words.
Yet by some it is said, the last speech that was made
Was so warm in that cause we admire
So the Broom[e] we must blame, for he kindled the flame
For *his words* set the house all on fire.

Says the Abbey, oh, dear! Such things I did hear,
I am sure 'twas a shame and disgrace,
For the rabble's vile cries, at each flame that did rise,
Made me blush, for I was red in the face,
One said, I declare, the Reform's in the air
And the Temperance Bill is up-raise'd
Hume's motion is won, by a majority of one
And the Poor-Law is all in a blaze.
But my friend, we can boast, we have long stood our post
Though the times are much alter'd, alas!
When we were built, I've heard men had a penny a day,

[d] Dorothy Jordan and William IV were Munster's parents. 'Jordan' was also slang for a chamber pot.
[e] Lord Brougham, Lord Chancellor and principal orator during the passage of the Great Reform Bill.

Guy Faux has rose again

And no law of oppression could pass.
Poor St. Stephen's is gone, its walls are forlorn
But, my friend, I now see day is dawning
It looks like a curse, but I'm glad it's no worse,
So I wish you a very good morning!

15

Friday, 17 October 1834   6 a.m.

Past peril

A T FIRST LIGHT on the Friday morning, spectators gathered in front of the Palace again to assess the extent of the devastation. The damage was terrible, and initially there was considered to be 'nothing striking, nothing picturesque' in the scene, despite contemporary romantic notions about gothic ruins. The firemen had relaxed their intensive efforts, but continued to pour streams of water on the smouldering embers for hours afterwards, as a precaution. There were at least three outbreaks of fire from the ruins that day. One, close to the south end of Westminster Hall at 8 a.m., was alarming but immediately subdued. The was another at noon and a further flare-up at six that evening, presaged by billowing clouds of smoke from above the southern gable wall of Westminster Hall. Firemen who had knocked off work to go to the Star and Garter were called away from their supper to attend it, where the wood had rekindled in the north-west corner of the Lords. Large plumes of smoke continued to rise into the air above Westminster Hall throughout the day.[1]

230

The approaches to New Palace Yard from Westminster Bridge were kept clear on each side by strong bodies of soldiers and police from daybreak. No one could pass through the barriers unless they were engaged in guarding the ruined buildings. Lord Broughton superintended various safety operations around the ruins for the best part of the day for the Office of Woods and Forests, and gave instructions to the firemen and workmen about the public property hastily dumped in St Margaret's Churchyard and the Speaker's garden. In both these locations soldiers were kept marching up and down for several hours and not withdrawn until the property was removed to safety. Books and furniture remained for many hours piled up in confusion under carpets and tarpaulins. It was a fine morning, and only a small shower of rain at two in the afternoon which dispersed the crowd 'helter-skelter' distracted from the melancholy mood. Before midday, a strong and extensive wooden hoarding began to be erected in Old Palace Yard to protect passers-by from walls which were in danger of collapse.[2] It was a wise decision. During the day, as the bricks and stonework cooled, 'considerable portions' of the ruins tumbled down or slumped to the ground, including a number of chimney stacks.[3] With each collapse, volumes of 'smoky dust' rose into the air.[4]

Melbourne and Duncannon returned to the Palace early, as well as Joseph Hume and some other MPs. Lord Althorp, the Marquess of Worcester, and even the gout-ridden politician and man of letters Lord Holland came to view the scene later in the day. Junior members of the royal family including the Earl of Munster also visited.[5] Crowds of sightseers were kept back by the police as the great and the good surveyed the destruction before them.[6] Thomas Creevy, the MP and diarist, wrote of how his half-brother, the Earl of Sefton, 'came up to-day on purpose to see the smoking remains of the two Houses of Parliament. What an event! I saw the poor old House of Commons smoking as I came over Westminster Bridge just now. The fire burst out again to-day.'[7] Sir Thomas Phillipps could be seen that morning occupied with wrenching out the papers that had got jammed between the cobbles of the road during the night, this time with William Black of the Record Commission helping him.[8] (It was an impossible task to attempt to retrieve every stray item: one unsubstantiated story from that time tells how some stray documents ended up as packaging in a butter merchant's on the Walworth Road.)[9] A number of Lords clerks—including Stone Smith—were already anxiously superintending the return of the books and papers to the Parliament Office.[10] The largely undamaged Royal Gallery was soon stacked with an incongruous array of hundreds of pieces of salvaged furniture from other parts of the Palace, including writing tables, chairs, drawers, steps, fenders, wardrobes, a clothes horse, several bedsteads, two chamberpot cupboards, a bidet-commode, and a thermometer in a mahogany case, 'broke'.[11]

The Speaker arrived from Brighton around twelve or thirteen hours after the fire and began to survey his ruined house and property, in despair.[12] Attended by several of the officers of both Houses, he was occupied for a considerable time during the morning in examining the ruins and ascertaining the extent of the damage.[13] With him was his son, and they anxiously toured the garden together, minutely examining drawers and desks for official papers.[14] His own furniture still being outside, and Mr Ley's house completely destroyed, Manners-Sutton wrote to the King at a table in John Rickman's house, no doubt requesting an audience at the earliest opportunity.[15]

All the Rickmans' valuables which had been removed by zealous friends found their way home that day, 'the arrangement of the Police and Military having been completed before dark—and the fire permitting little lurking Roguery in its glare'. John Rickman was delighted to report to friends, 'I seldom acquire *exact* information till I wish to apply it to use—and as I have not office or room at the Commons—not a closet or Drawer except that in the Table—I have lost nothing there.'[16] On unpacking, Frances had the dubious pleasure of discovering all her gauze dress bows squashed under her heavy outdoor clothes.[17] At nine on the Friday evening several of the fire engines departed, but a fatigue company of the Guards was posted beside those which remained on the ground in case of any further problems, and stayed throughout the night. A small crowd was still there, but the authorities now took advantage of the lull to erect two further barriers—at the Canning statue to the north and at Abingdon Street to the south. By then hoarding in front of the Lords was nearly complete.[18]

On Saturday, 18 October, officials of the House of Lords continued to be occupied in superintending the return of books belonging to their Library. Some objects had been removed to the homes of Parliamentary clerks who lived nearby (some 'as far off' as Millbank) and 'have sustained much less injury than, under the circumstances of so unceremonious, confused, and rough a removal, might have been expected; now is it ascertained that there have been very serious losses, dreadful, even frightful, was the plunder in other directions'.[19] Firemen were now busily engaged in pumping the water from the coal cellars beneath the Commons Members' Entrance opposite St Margaret's while others were bringing up yet more books and papers completely saturated with water.[20] Outside, the nine-foot high hoarding was finished, blocking from view most of the ruins. Workmen were clearing away 'and carefully examining the rubbish in all those parts where the heat has sufficiently evaporated to allow them to be handled'. The police remained in attendance to prevent looting.[21]

Lady Manners-Sutton arrived by coach from Brighton at two o'clock on the Saturday, with her daughter. Supported by her son and Mr Fairlie, she entered the hall and immediately burst into tears at what she saw. She was taken over to the staircase by her daughter and Fairlie, where she greeted two of her housemaids, who had saved her dresses, china, and valuable articles of decorative furniture. Shaking their hands and thanking them, she then recovered herself enough to tour the house, though still 'with feelings of horror'.[22]

Meanwhile, Lord Broughton had been to St James's Palace and had been granted an audience with the King, who had arrived in town from Windsor with the Queen that morning. The topic, of course, was the fire. 'I cannot say he was much affected by the calamity,' noted the First

Commissioner of Woods and Forest drily, 'rather the reverse.'[23] The Speaker also waited on the King and, after a long conversation about the catastrophe, was about to take his leave when the King said, 'I suppose you are now going home, Sir Charles?' The Speaker was nonplussed by the question and the King recovered himself sufficiently to say, 'I beg your pardon Sir Charles, I had forgotten you have not got a home now—I mean are you going to Palace Yard?' When Manners-Sutton said he was, the King decided to go too, 'in as private a manner as possible'.[24] Accordingly, at three o'clock that afternoon, William and Adelaide drove to the Palace of Westminster in two plain carriages and toured the ruins of the two Houses. The party included the Earl and Countess of Errol, William's three sons who had been at the fire—the Earl of Munster, Lords Adolphus and Frederick FitzClarence—and several other noblemen. A crowd of people pressed in and went round with them.[25]

Among that crowd were John Rickman, Frances Rickman, Sir Francis Palgrave, and his 10-year-old son Frank (later the editor of the bestselling poetry anthology *Palgrave's Golden Treasury*). Frances Rickman recalled how the King and Queen arrived privately, soon after they had gathered, in a surreal version of the usual royal visit at State Opening:

> they approached us in procession just at Mr Ley's end and the Speaker shook hands and spoke to us and said we might go with them which Sir F, Frank and I did—I was next Mr Repton and spoke to him and he knew me and said if I would keep by him we should see all. It was a melancholy and curious sight—the King in plain clothes, the Queen in a white satin bonnet...and a dark merino cloak—I had the honor of being the only lady present!

Some men were present with a piece of tattered red carpet and matting, which was all they had to hand, and were laying it first before the royal feet in one spot, then once William and Adelaide had passed, taking it up and using it again further down the route. Frances found it poignant to walk through the remains of the Royal Gallery entrance where, on happier occasions, she and her sister had strained their necks to see the royal couple when the bells and the cannon announced their arrival at the Palace.[26]

On viewing the ruins, William was stunned at what he found. He was reported to have been fixed to the spot for some time, without apparently having the power to make any observation: 'amazed at the fury of the flames which could have hurled so much destruction through and over these enormous buildings and, in many instances, terrifically thick old stone walls'. Eventually, he and the Speaker moved on to look at the state dining room under St Stephen's, where the King remained for a long time, viewing every detail with 'painful anxiety' and in close conversation with Manners-Sutton, his characteristic 'quickness of observation and inquiry' failing him. Whatever his true feelings earlier in the day, *The Examiner* felt that in the afternoon visit he clearly demonstrated 'one painful feeling of afflicting thought on the ruin of so politically sacred a spot'. Adelaide was equally concerned and interested.[27] Broughton worried during the tour that the King and Queen 'went rather too near to the shaking walls, but they were quite fearless'. He was not so sure about the King's mood, thinking William 'looked gratified as if at a show', and was told by the King, 'Mind, I mean Buckingham Palace as a permanent gift! Mind that!' After a backwards glance as the King was leaving, the royal party returned to St James's and thence swiftly back to Windsor.[28]

On Sunday morning, 19 October, St Margaret's Church was still so stacked up with salvaged records that worshippers hoping to attend an advertised charity sermon in the morning were diverted to St John's, Smith Square.[29] Frances Rickman went twice to St John's to give thanks for the family's deliverance, once to Mr Williamson's charity sermon and then to Mr James's service in the afternoon.[30] After church, tens of thousands of people from across London spent their day of rest gawping at the ruins from a distance. The crowd became so packed behind the barrier at the northern end of Westminster Hall that the police began to allow people to enter through an opening, and from that time onwards, spectators were permitted to promenade within the cordon to the top of Abingdon Street, and then back again, providing they did not stand and stare. Smoke was still rising from parts of the building, and window frames would every now and then break out into small flames. Several more walls fell down during the course of the morning. Rough signs had been placed on the buildings painted with the warning BEWAIR OF DANGER. The spelling was adversely commented on.[31]

On Monday, 20 October, several pictures were finally returned to the Palace by the owners of nearby houses where they had been placed for safekeeping.[32] Following the royal visit to the site, John Rickman had promised his family that they would go on a more leisurely tour of the ruins in the next few days. 'Our feet will be warmed, I expect,' responded his wife.[33] Accordingly, they viewed the ruins on the morning of Tuesday, 21 October, once the house had been straightened out, everything more or less put back in the right place, and their frayed nerves were a little calmer. The sight shook Mrs Rickman a little, and she was glad to turn her steps homeward again with their party, which once again included Sir Francis Palgrave—this time with Lady Palgrave, sister Hannah, Mr Wood, Miss Sexton, their friends the Wildes, and Mr Gurtkin. Some beautiful parts of the old stonework remained, she noticed, and sketchers were positioned 'at favourable points to give them to posterity'.[34] The London Fire Engine Establishment withdrew its last engines damping down the embers later that day.[35]

The day after, 22 October, Susannah Rickman walked around the district, calling on friends to thank them for their kindness in receiving all their removals, and in making offers of beds. A young man was taking sketches inside the ruins. She could not resist returning home without taking a peep from St Margaret's Churchyard at the 'fine old walls of the House of Lords, now completely laid open by the entire fall of the Cotton Mill screen which surrounded it (built on as a lean-to Aunt L describes)'. Some which remained had already been pulled down. As she watched, a rope was fixed at the top of a remaining tower in the corner overlooking Old Palace Yard, at which the crowds were gazing. She did not stay to see it fall. Above all, she was thankful for the preservation of her family and their home, next door to the building whose ill-advised clearance had set in train the catastrophe. 'Do you not admire the dutiful Tallies?' she wrote to her daughter, Anne. 'They would not singe their old Master the Exchequer!'[36]

Investigations at the highest level began almost immediately. There was an emergency Cabinet meeting on the Friday, at the Home Office, of those Ministers who were in town.[37] Between 17 and 19 October, witness lists had been drawn up, initial statements taken, and preparations made for a formal inquiry.[38] On Monday, 20 October, the Privy Council began to hear evidence on the origin and cause of the fire, amid swirling rumours. Some in the know said it had begun in the roof of Howard's coffee-house, occasioned by experiments with some new stoves; that it broke out in a passage leading to the bar of the House of Lords; that it was an overheated flue in the Stranger's Gallery. Some, including Mr Bellamy the caterer, asserted that it had begun right in the middle of the Lords Chamber.[39] Others heard that it had originated in Mrs Wright's room. Speculation as to its cause ran from the general (a gas explosion) to the more specific (the bursting of 'some of the gas tubes which pass through the interior of the House' or at the Peers' Entrance). Another story blamed 'the carelessness of some of the servants at Howard's coffee-house, where a dinner had been given to a party of gentlemen and the cook had incautiously thrown some combustible matter into the stove, where some live embers still remained', and yet more tales fixated on the actions of plumbers in the Lords Library, or 'some workmen, who had been engaged in doing some repairs in the body of the House had, on leaving work, left a parcel of shavings near a grate, in which were the remains of a fire which they had been using in the course of their work'.[40] Government cutbacks were also implicated. It was alleged that the man 'whose exclusive duty it was to be in constant attendance at the House to watch the fires' was now impossibly stretched to do the same at St James's and

Kensington Palace as well.[41] Trying to squeeze the new Court of Bankruptcy into the old Exchequer offices, instead of building enough accommodation in the Law Courts in the first place, was also highlighted: 'Economy has cost the country three hundred thousand pounds.'[42]

Elsewhere, more sinister forces were thought to be involved. Lord Melbourne received an anonymous letter suggesting the fire had been started by 'an operative' paid £30, who had thrown turpentine against the walls of the House of Lords to increase the blaze, and had 'deliberated whether it should be set on fire now or when the members were sitting'.[43] It was whispered that some in the crowd on the night were heard to say, 'the union men did this'.[44] Other tales 'of an injurious tendency were circulated, but no shadow of proof was offered'.[45] Yet even before the official inquiry sat, some of the newspapers had been conducting their own investigations. On the morning of 17 October, *The Times* was already getting warm by asserting that the probable point of origin were 'the flues which have lately been repaired, and in which some experiments have been making for the purpose of more efficiently warming the House of Lords'. As to its cause, however, that question 'is of necessity at present involved in mystery, owing to the extreme confusions and bustle that prevailed'.[46] By the day after, *The Times* had nailed it. Tallies burned to heat the flues were the cause, it declared, with a suggestion that the burning might even have been undertaken to destroy the tallies themselves, and so, 'the quick heat produced by the destruction of the old tallies, and some unknown defect or foulness in the flues, are supposed to have been the means of igniting some of the surrounding timbers'. The strictest inquiry was in progress, it added, 'but there is not the slightest reason to suppose that it has arisen from any other than accidental causes'.[47] The following day, 19 October, *The Examiner* pronounced with confidence the 'general opinion of all in authority that it originated in the burning of the tallies'.[48] Some people, used to opining freely on current affairs, felt it was pointless, given that the fire was the only topic of conversation everywhere, and the newspapers were filled with such a 'god-send of a subject'.[49]

The diarist Greville wrote, somewhat petulantly, 'for two months nearly that I have been in the country I have not written a line, having had nothing worth recording to put down...it would have been a mere waste of time to copy the accounts of the conflagration of the Houses of Parliament'.[50]

The inquiry itself excited 'the most unbounded interest'.[51] Of the twenty-six members of the Privy Council, the hearings attracted an unprecedented twenty-one to attend, including Melbourne, Duncannon, and the Speaker himself. It was intended that they should be held in conditions of the utmost privacy, and even Mr Gurney, the official shorthand writer to both Houses transcribing proceedings, was sworn to secrecy. Not surprisingly though, somehow the evidence of each witness leaked out daily, resulting in reports in the papers which managed to be both detailed and garbled.[52] The Lord Chancellor, Brougham, examined the witnesses.[53] (Brougham, incidentally, was described as a 'two-fold sufferer'. In a rather tall story, his House of Lords wig was said to have perished, as well as his wig in the Court of Chancery.[54]) The Attorney-General, Lord Campbell, who was also present, was aware that many outside had a 'great desire to make out a conspiracy', but 'the case was clear from the beginning, and the investigation is only continued for public satisfaction'. The knee he had damaged pumping at the fire was still giving him gyp during the first week of hearings.[55] In this febrile atmosphere, and three days into the inquiry, it was reported that Windsor Castle had burnt down.[56] The witnesses were variously terrified, shocked, or determined to blame someone else as they were questioned. Joshua Cross gave his evidence in between shifts working at the ruins.[57] A substantial amount of the inquiry's time was occupied in investigating claims by a Mr James Minion Cooper, owner of Hall & Cooper, ironfounders of Drury Lane, that he had heard news of the fire at an unfeasibly early hour in Dudley in the West Midlands. After exhaustive questioning, the inquiry came to the conclusion that Cooper must simply have mistaken the time.[58] Cooper's motivation to circulate such a story may have been to advertise his business—significantly, a manufactory of the Marquis de Chabannes' Patent Apparatus for warming and ventilating buildings.[59] If so, it failed, merely distracting the Privy Council from its investigations into the true cause, and delaying proceedings.

On 8 November, the *Report of the Lords of the Council respecting the Destruction by Fire of the Two Houses of Parliament* was published along with extensive minutes of evidence. Its findings were that the uncontrolled burning of the tallies was to blame for the fire; that it was unfortunate that Weobley did not supervise the burning better; that Cross and Furlong were guilty not of malice, but 'gross neglect, disobedience of orders and utter disregard of all warnings'; and that it was unaccountable that Mrs Wright did not make representations to the House authorities as soon as she became alarmed. The Privy Council concluded that 'the fire was accidental, was caused as we have related it, and was wholly attributable to carelessness and negligence. Indeed it would be very difficult to point out a case of fire which could more clearly be traced than this has been to its cause, without suspicion of evil design.'[60] No one was punished.

'Oh dear, oh dear, what a consternation | This affair will cause throughout the Nation', rhymed the balladeer John Morgan.[61] It caused more than consternation. It set in train a series of events which changed Parliament and the country forever. The diarist and man-about-town Thomas Raikes, residing in Paris at the time, felt the catastrophe was ominous:

> The two contending bodies of the state, in dire opposition to each other, the one insolent and overbearing in aggression, the other strict and obstinate in defence of its privileges, both buried in one common ruin.... The old walls of St Stephen's have witnessed a long career of British Glory and prosperity; may it not have perished with them!!! Time will show that mystery. But if the character, talent, and honour of those public men who in years gone by have distinguished themselves within those walls contributed to support that career of glory, then may we own that they have now crumbled over the heads of men who are utterly incompetent and incapable of maintaining it.[62]

In November, while Parliament was not sitting, the King used the spu-
rious excuse of Lord Althorp's elevation on the death of his father Earl
Spencer to dismiss the government, Althorp having given up his position as Leader of the Commons. Wellington formed a caretaker administration until Peel could return from Rome, where he had been
throughout the fire crisis. Once Peel was installed as Prime Minister on
9 December, the King dissolved Parliament and called a general election. But the Commons would not have it, and Peel's shaky coalition
lasted only until April 1835, when Melbourne returned to power. It was
the last time a British monarch exercised their constitutional right to
dissolve Parliament. The world had indeed changed.

On 23 October, at two in the afternoon, the new session of Parliament opened in the House of Lords Library as planned. Some salvaged books had been replaced on the shelves 'in the most irregular
manner', and a gold-burnished chair was placed at the end to represent
the throne of George IV, which had been burnt to a cinder. In front of
it was a form for the Lords Commissioners to sit on (deputizing for the
King), and a miniature woolsack, with benches and cross benches covered in scarlet cloth in the usual arrangement found in the old Chamber. The original despatch boxes were present on the table of the
House. Ten peers were in attendance, including Lords Munster, Auckland, and Hill. Mr Courteney, the Clerk of the Parliaments, was 'taking
minutes as coolly as if nothing had happened', with Mr Currey, the
Lords' Reading Clerk, calmly reading the commission. Both were as
immaculately robed and wigged as usual. Black Rod went to demand
the attendance of the Commons from where they had assembled in the
unburnt Lords committee rooms nos. 4 and 5 beyond the Royal
Gallery.[63]

Mr Ley, the Clerk of the House, and his son, the second Clerk Assistant, were unable to play a part. The fire had deprived them of even
their most obvious possessions: without their wigs they could not participate. The Leys thought it a great joke.[64] Accordingly, once the Commons was called to attend the Lords, John Rickman deputized for Ley,

and processed to the makeshift Lords Chamber to hear the message with a few members and all the clerks and other officials. The message from the King, read out by Brougham, was that Parliament was immediately proroguing itself to meet again on Tuesday, 25 November.[65] That afternoon, a week to the day after the disaster began, selected visitors began to be admitted by ticket to the ruins. The old retainer George Wright, husband of Mrs Wright, was posted by the gate to oversee access to the site.[66] Coals from the cellar emitted fire at intervals and the British fire engine was obliged to be put into action.[67] That evening the Rickman family gave a dinner to 'glory in past labours and past peril' and to thank the friends and all those who had cleared their house during the terrifying hours while the fire was raging towards them.[68]

Epilogue

LTHOUGH THE FLAMES were out, the conflagration of 1834 continued burning for many years afterwards: through individual lives and throughout British social, cultural, and political life. It occurred just at that moment when the Georgian period was giving way to the Victorian; as the stagecoach was overtaken by the railway; as the ancient parish system was swept away by new metropolitan services; and at a time when the last gasps of the medieval city and the first breaths of a modern one were being heard.

No one died in the fire, despite tales to the contrary. But at three o'clock on the afternoon of 23 October 1834, exactly one week after it began, an elderly labourer called Edward Saunders was lowering a rope for wood when he lost his balance, fell off his joist, and plunged fifteen feet to the ground. He was known as 'a steady man' who had been part of the team which had erected the cast iron at Buckingham Palace, and lived at 10 Dollop Court, Westminster. He died on his way to Westminster Hospital from a smashed skull and broken neck.[1] Had he reached the Hospital, he would have discovered that four out of the nine men severely injured in the fire were still there. Fireman Hamble-don was the greatest sufferer. His ultimate fate is unknown; the others were discharged a few days later.[2] On Saturday, 1 November, John Bow, another labourer, aged 45, was carting away rubbish from the ruins, working in a hole about three feet from an unstable wall near the shell of the Commons Library, when it collapsed on top of him and he was retrieved—crushed to death—two hours later. The wall had previously been sold to Hickmold, Bell & Riley, a salvage company, and its sup-ports had been removed following its purchase. Hickmold & Co. was ordered to contribute some compensation to Bow's widow and her four children, living at 9 Little Chester Street, Belgrave Square.[3] Another fire widow, Mrs Hannah Cogger, was paid £5 in compensation follow-ing Mr Cogger's death 'occasioned by cold taken at the Fire'; her hus-band having been one of those aiding the rescue efforts on the night.[4]

The twisted knee of George Manby, erstwhile inventor of the fire *extincteur*, kept its owner inside for several days. Once mobile again, Manby went to see Colonel Rowan, the Police Commissioner, who ex-pressed the view that there had been 'a want of union among persons in the direction of Fire engines' and provided Manby with a police escort to view the ruins of the Palace. There he retrieved, and gave to Dawson Turner, a charred fragment of a Commons order paper, which survives in the Library at Trinity College, Cambridge.[5] The Commit-tee of the London Fire Engine Establishment wrote to the Duke of Wellington in early December, to point out that the private insurance companies making up the Establishment were still duty bound to their private employers and, 'if during the late conflagration at Westminster, any insured property in danger—or any simultaneous fire or fire in other parts of the town, had imperatively called upon the Superintend-ent to devote the Services of the Engines elsewhere, Westminster Hall and the Public property adjoining must have shared the fate of the Houses of Parliament'. They recommended that the problem 'would be corrected merely by placing the parochial engines under the inspec-tion of the Commissioners of Police...and by placing the public and parochial engines *at fires* under the order of *one* directing officer'.[6] The papers also strongly supported the establishment of a single fire super-intendent, and even the institution of a Fire Police, to take charge of public safety on such occasions, 'with one ostensible and responsible head'.[7] The police themselves had been found to be 'firm, temperate and judicious; no number of parish constables could have performed their duty, and ten times the number of soldiers would have been less effective'.[8]

'Though I regret the extent of the fire and the loss which has been sustained by the country,' James Braidwood had reported to the LFEE board, 'I have the satisfaction of feeling that no exertions were spared, either by myself or the men under me, to arrest the progress of the flames. I also gratefully acknowledge the very efficient assistance of the military in working the engines and of the military and police, in keeping the ground clear.'[9] This comment was typical of Braidwood, a devout member of the Church of Scotland, kind, modest, adored by his men, and quite brilliant in his creation of modern firefighting theory. Typical too was his immense bravery—he was to die, as he had lived, fighting fires, at the enormous Tooley Street warehouse fire in 1861, when he was buried under a falling wall, and where the tallow inside the warehouses set fire to the Thames for two weeks afterwards. The iron firedoors which he had designed and recommended for all such premises had been left open, causing his death. Mourners at his funeral stretched for one and a half miles following his coffin.[10] A single London Metropolitan Fire Brigade was finally established in 1866, after continual pressure to form a unified service, beginning with the 1834 fire.

Within forty-eight hours, artists were climbing on the ruins of the House of Commons to take sketches.[11] For weeks afterwards the ruins were 'thronged with artists, making drawings of the ruins, at no little risk of personal injury… since the late rains, the stones have tumbled all around them; and the falling mortar has given many an unintended tint to the sketches before them'. The martyrdom of St Stephen—through stoning to death—was recalled, somewhat heavily, but overall the mainly dry weather helped the artists.[12] At least forty-two artists are known to have produced scenes of the fire, or of the remains, including the younger son and sister of Charles Barry and the splendidly named Hercules Brabazon Brabazon. Many of these works, which are often the only surviving evidence of the structure and fabric of the old Palace, have been acquired over the years for the Parliamentary Works of Art Collection. Of particular note are those of 21-year-old Robert Billings, who produced many watercolours documenting the damage in detail and who had been apprenticed to the topographical draughtsman John Britton.[13] In fact, on 22 October, Susannah Rickman had noticed 'a young man from Mr B's' making beautiful sketches within the ruins.[14] Thirteen-year-old George Scharf, later eminent Director of the National Gallery, accompanied his artist father to the scene, where he watched his father prepare a spectacular panoramic diptych in oil from the roof overlooking St Stephen's. Lost for many years, one panel was recovered a few years ago, and is now in the Parliamentary Works of Art collection.[15] Two antiquarian publications emerged from the wreckage. Edward Wedlake Brayley and John Britton's *The history of the ancient palace and late Houses of Parliament at Westminster* was published in 1836, a book in fact begun before the fire but whose subsequent topicality made it walk off the shelves when published. This was followed by Sir Frederick MacKenzie's romantic reconstructions of *The architectural antiquities of the collegiate chapel of St. Stephen, Westminster* in 1844. Less elevated productions were available much more quickly. Within days the *Mirror* was offering back copies of its post-fire editions, containing illustrations of the west front burning, the ruins, and even an engraving of a tally.[16] A print of the *Houses of Parliament on FIRE* was being distributed within a fortnight, costing 9*d.* on coloured tinted paper or 1*s.* 6*d.* in colours printed on cardboard.[17] It was followed by many more, including a cartoon portraying John Bull declaring to some peers, 'It was a great pity when the filthy old rookery was burning that the rooks warn't in it at the same time', while Brougham removes the burnt remains of the Woolsack.[18]

Curious visitors got somewhat out of control at the ruins and became, to antiquaries' eyes, 'public enemies'. The beautiful upper tier of the chapter house in the cloisters suffered 'much wanton destruction from the disgraceful cupidity of the relic hunters, who have mutilated its exquisite niche-work... This is truly lamentable; and is enough to make one curse the whole tribe of collectors, whose senseless mischief has destroyed what the fire had left perfectly uninjured.'[19] Other relic hunters could make their way to Mr Doubleday's in Little Russell Street near the British Museum. There, two kinds of souvenir of the old Palace were on sale. The first were oak boxes made from a beam under the Painted Chamber, each inset with a brass escutcheon made from a melted brass chandelier from the House of Lords. The discerning could also obtain an oval seal cast from lead allegedly from St Stephen's roof, and displaying the figure of St Stephen, with canons of the college praying before him and the Virgin Mary and St John.[20]

The precedent of allowing artists and tourists to view the ruins of a disaster was set at Westminster 1834. At the Tower of London, almost exactly seven years later, a serious fire destroyed the Armoury buildings. When that occurred, the authorities responded differently. The site was closed, and then following a clear-up reopened a month later, charging sixpence per visitor to tour a pre-determined route, with carefully arranged exhibits of the damage. They even opened a souvenir kiosk where remnants of the fire-twisted metals and other destroyed items were on sale to curious relic hunters.[21] It is obvious that in 1841 the Tower authorities had learnt from the Westminster experience, and had decided to cash in on the visitors' interest themselves, rather than letting private entrepreneurs (or royal servants) take all the profits, and risk damage to the fabric which needed to be protected.

Within the week, a moving panorama of the fire was on display at the Victoria Theatre, created by the scene-painter Charles Marshall, the inventor of limelight for the stage.[22] At the Adelphi Theatre an exhibition of the scene, as viewed from the Lambeth bank of the Thames, was mounted in competition, and was thought to have succeeded well in portraying the Palace, Westminster Bridge, and the crowds on land and water. Some critics complained that the fire appeared to be behind the buildings, rather than inside them, and the smoke was rather sparse—but the intense heat in the stuffy venue was certainly considered to be highly realistic.[23]

Incredibly, a few weeks after the fire, one of the Armada Tapestries was discovered safe. Mr Thorn, a dealer in antiquities and curiosities, had purchased it in May or June 1834 from a furniture broker called Preston, of Clare Market, for 30 shillings. It was about 27 feet (8m) long and 15 feet (4.5m) deep—and in a 'very decayed state, very much worn and almost reduced to threads'. There were allegations it had been used as a carpet. Attached to it was a label stating it was 'the piece opposite the throne'. It turned out to have been removed from the Lords Chamber during the building of an additional gallery for the public wanting to watch the Great Reform Bill debates in December 1831. Packed away in a wooden chest over a closet by a German servant of the Lord Great Chamberlain's secretary, it had been considered perfectly safe by his master. On his retirement in early June 1834, the German, a man called Seiben, had offered it to a porter at the House of Lords called Ware for five shillings, without realizing what it was: 'My master had never given me this carpet, nor any permission to take it...my belief was that it was an old piece of furniture of no value...Nobody is to blame. It is entirely my own fault.' Now, with a potential goldmine on his hands, Thorn approached Wellington pointing out that,

> the late conflagration having destroyed the corresponding tapestry, I conceive it has now become an interesting relic; and although it has, in consequence, become of great value to me in my business, and I should be most proud still to preserve it as an object both of curiosity and profit, being adapted, after repair for public exhibition; yet conceiving it to be an ancient record of national history, I feel it my duty most humbly to represent the fact to your lordships; and if his Majesty or your lordships should think it of sufficient importance to merit your attention, I will cheerfully attend your lordships' commands when and where your lordships may be pleased to appoint, relying on your

lordships' liberality to remunerate me for the preservation of so valuable a remnant.[24]

Thorn now believed that £400 might not be an unreasonable sum.[25] The government refused to purchase it. 'This object', the Lord Great Chamberlain was advised, 'is of no sort of worth, the recent destructive fire having alone given adventitious value to it… it could never be applied to any purpose, either of ornament or convenience.'[26] In the end Edward Copleston, bishop of Llandaff, bought it in February 1835 and then donated it to the City of Plymouth as the piece depicted the battle there ('a judicious and most appropriate disposition of it', concluded the Lord Great Chamberlain). It was housed in the Plymouth Institution (later the Plymouth Athenaeum), which disposed of it in the 1880s. Its significance by then seems to have been entirely forgotten.[27]

A Royal Commission was set up in 1841 to make recommendations on the artwork for the new Palace of Westminster. Chaired by Prince Albert, it chose a Tudor theme for the Prince's Chamber, the ceremonial anteroom adjacent to the new House of Lords Chamber. Its threefold scheme comprised a series of portraits of Tudor kings and queens, some bronze relief sculptures, and, for the upper tier of the double-height room, a cycle of paintings reproducing six of the Armada Tapestries. The portraits and bronzes were completed in the 1850s, but the death of the Prince Consort in 1861, technical problems, and dwindling funds meant that the tapestry reproductions fell by the wayside, apart from a demonstration canvas executed by the artist Richard Burchett. The six empty panels were then covered over with Pugin wallpaper. In 1907, an attempt was made to revive the plan, but it foundered. Then in 2007 the House of Lords Works of Art Committee approved a proposal to complete the scheme, funded by a generous donation by the American philanthropist Mark Pigott. A monumental two-year project ensued, in which the artist Anthony Oakshett, aided by five other artists, re-created the five remaining canvases (each one roughly 12 feet (3.7m) by 14 feet (4.3m)) based on Burchett's initial sample, digital blow-ups of John Pine's 1739 engravings, and the input of many historians and curators. These magnificent paintings were installed in the Prince's Chamber in autumn 2010, where today they can be admired by visitors touring the Palace, and not too far from the site where the original Armada Tapestries hung until 1834.[28]

Based on the two sketchbooks he had filled on the night of the fire, Turner produced one large watercolour of the view from Abingdon Street (today at Tate Britain), a vignette, and two huge oil paintings from the Lambeth bank and from Waterloo Bridge downriver. Well known for completing his paintings on 'varnishing days' (that is, once the half-finished canvases were hung in the galleries, prior to their official opening), in February 1835 Turner finished one of the oils of the *Burning of the Houses of Parliament* in public at the British Institution, and probably did the same later that year with the other version, at the Royal Academy. Critics at the time were somewhat mealy-mouthed about them, puzzled and even dismayed by the gulf between Turner's two portrayals of the fire and their memory of it. They wanted to see something more representational, not something where 'truth is sacrificed for effect' and where one had to be a certain distance away to appreciate its full power. Turner's pictures seemed a 'splendid impossibility' to the *Athenaeum* critic, who complained about the colour of the sky and declared one of the pictures to be a 'heap of huge daubs'. Today those two pictures—now at The Cleveland Museum of Art and the Philadelphia Museum of Art in the United States—are regarded as among the crowning achievements of Britain's greatest artist.[29]

Chance, that 'extraordinary specimen of canine eccentricity', died just before the first anniversary of the fire. Obituaries appeared in the national and provincial papers, reporting that even as he lay dying he attempted to go to one last fire, but collapsed at the door. The twelve LFEE firemen of the Watling Street HQ each subscribed a sovereign to have him stuffed 'by a regular professor' and placed in a plate-glass case, three feet six inches long and two feet two inches wide, with an enamel-gilt border. A scroll above commemorated

Chance, well known as the firemen's dog, died October 10, 1835. This is
humbly inscribed to the Committee of the London Fire Establishment,
by their obedient servants.

The names of the contributing firemen were listed on either side of the
memorial. The return of the immortalized Chance to Watling Street
was delayed when the taxidermist decided to sell him on to a showman
who insensitively exhibited him in his case in a distant part of town,
where crowds flocked to see him at a penny a head. The ruse was uncovered when a disguised fireman paid his entrance fee and discovered
the sacrilege. A body of firemen subsequently retrieved their dog in
force (despite being offered 12 guineas to part with him legally), and
reverently placed his glass case on a table at the back of the long room
in the fire station. A collecting box stood nearby, and a wag wrote in
chalk on the lid of it, 'It is all a chance'. Chance apparently left a son
and heir 'which was made quite a pet of, till some *fancier* took a fancy to
it, and stole it'.[30] Chance's whereabouts, and that of his glass tomb, are
today unknown.

New chamber accommodation was needed for the two Houses. With
the prorogation of Parliament, the Office of Woods and Forests had
obtained a month's breathing space in which to find a solution. Suggestions sprang up thick and fast: Westminster Hall; the Guildhall in the
City of London (where the Commons had sat during the Civil War);
St James's Palace; St Margaret's Church; Marlborough House in Pall
Mall; Whitehall Chapel (that is, the Banqueting House); the Halls of
the Inner and Middle Temples; even a temporary structure in Old
Palace Yard, connecting with Westminster Hall and surviving offices,
was suggested.[31] One option—reroofing St Stephen's—was out of the
question. It could not possibly be stabilized, and it might even need to
be demolished.[32]

The King eagerly offered up Buckingham Palace.[33] The papers
became convinced that this was to be the chosen replacement. 'We
understand', the *Morning Chronicle* informed its readers, 'that upon receiving the intelligence of the destruction of the Houses of Parliament,

his Majesty, with that devotion to the public interest which has invariably marked his conduct, immediately placed the Palace newly erected in St James at the disposal of the nation, in order to meet the difficulty of the emergency. It is not yet determined whether the generous offer is to be accepted.'[34] On 24 October, *The Times* announced that the next Parliamentary session had been fixed for 'the new palace at Pimlico', and that work was already under way to stop the existing development, and to refit it for the two Chambers.[35] Measuring-up had begun, said the *Morning Chronicle*.[36]

The old arguments soon resurfaced. *The Examiner* (preferring St James's as the new venue) called Buckingham House a 'crude unsightly building—a disgrace to the age in which we live'.[37] And Lord Holland noted, 'The King, who in truth, dislikes the new Palace of Pimlico, lost no time in offering it… but he had not the King Craft to conceal that part of his motive was the indulgence of his own inclination and therefore got little credit for it.'[38] Melbourne was determined that William should not try to unload the embarrassment on to Parliament. The offer was tactfully turned down.

Yet behind the scenes, at first scarcely noticed by the papers, another plan was rapidly being hatched by the Office of Woods and Forests. Robert Smirke had been commissioned to undertake a survey of the damage in the Palace of Westminster. He had reported on 21 October that with some repair, the ancient walls both of the Painted Chamber, which were badly fractured, and of the Lesser Hall would nevertheless be sound enough 'to receive a Roof and the fittings necessary for their temporary use with perfect safety'. The Lords could move into the Painted Chamber, which connected to the mostly undamaged sections of the Lords—the Royal Gallery, committee rooms, and Library. The Commons could sit in the former Lords Chamber, and some of their damaged committee rooms repaired. The alternative of fitting out Westminster Hall with two Chambers was not necessary.

The repairs undertaken as a result of the fire provided the first opportunity to allow women into the Commons to view proceedings officially, with twenty-four seats being provided in a dedicated Ladies' Gallery, behind a trellised screen and up a separate staircase.[39]

The revamped Chambers were ready for use less than three months later (Parliament having been prorogued a further twice in the interim to allow repairs to continue).[40] This was an incredible feat, achieved by the contractors S. Baker & Son, T. & G. Martyr, and Charles F. Bielefeld through the use of night-shifts, prefabricated timber and iron girders, and papier-mâché ornamental mouldings.[41] It cost £32,140 to refit both Chambers and provide temporary committee rooms (including one offsite at 30 Great George Street); build new refreshment and prison rooms; fit out an office for the sale of official publications in Abingdon Street; and to improve the Speaker's accommodation, which had been transferred to Carlton Gardens on the other side of St James's Park. Numbers 43 and 44 Parliament Street were requisitioned to provide offices for the Commons administration. The cost of refurnishing came to £13,441.[42] The Lords immediately began complaining that the Painted Chamber was too small for them; while members of the Commons were so delighted with their spacious new accommodation in the former Lords that they became quite fond of it and complained loudly about the ventilation and acoustics of Barry's new Chamber when they first moved into it in 1850.[43]

At noon on 19 February 1835, the doors of the temporary House of Commons were thrown open and three hundred members swarmed into their new home. There they waited for two hours until summoned to the Lords by Black Rod to hear from the Lords Commissioners. 'At no period of this country's history, under no circumstances, public or otherwise, that I can conceive, could a fitting selection of an individual from among us, to fill the Chair, be a matter of greater public concern, or a Motion which tends to that selection a question of more serious responsibility,' said Francis Egerton MP, who proposed Manners-Sutton once more as Speaker. But it was not to be—the former Tory Speaker was not voted back again. Instead, the Whig James Abercromby took his place, in a signal to the King of the Commons' disapproval of his peremptory dissolution in November. Over the next few days, new Members were sworn in.[44]

William IV opened Parliament in state on 24 February, processing to the Painted Chamber in an echo of the Parliamentary gatherings of the thirteenth century. 'You will, I am confident,' he pronounced to the Lords and Commons there assembled, 'fully participate in the regret which I feel at the destruction, by accidental fire, of that part of the ancient Palace of Westminster, which has been long appropriated to the use of the two Houses of Parliament. Upon the occurrence of this calamity I gave immediate directions that the best provision of which the circumstances of the case would admit, should be made for your present meeting; and it will be my wish to adopt such plans for your permanent accommodation as shall be deemed, on your joint consideration, to be the most fitting and convenient.' The King then directed that the Privy Council's report on the origin of the fire would be the first paper laid on the table of each House that session.[45]

Frances Brandish was the last House of Lords Housekeeper sinecurist. When she died in 1847, the Housekeeper's post was turned into a directly salaried post of the House of Lords. The new postholder was a Mrs Bennett—in fact, Jane Julia Wright, who had remarried.[46]

Richard Weobley remained in post as Clerk of Works at Westminster until he was transferred as Clerk of Works at the Brighton Pavilion in August 1840, in an act that was no less than Siberian exile by that date.[47] He was a broken man. Corruption in the Office of Works was rare at the time, but in the most serious case of its kind, he was finally dismissed on 25 June 1847 for selling old lead and brass belonging to the Office, and lining his own pocket with the proceeds.[48]

Aged 40 at the time of the fire, the volatile Earl of Munster sank into depression on the accession of his legitimate cousin Victoria in 1837, and committed suicide in 1842.[49]

Charles Dickens published his first novel, *The Pickwick Papers*, in episodic form, beginning in 1836. He gave up work as a Parliamentary reporter, but developed a lifelong interest in political reform, reminiscing on the fire (and commenting on the cost of the new Palace) in later life at one of his immensely popular public lectures on administrative reform:

> There is, however, an old indisputable, very well known story, which has so pointed a moral at the end of it that I will substitute it for a new case: by doing of which I may avoid, I hope, the sacred wrath of St. Stephen's. Ages ago a savage mode of keeping accounts on notched sticks was introduced into the Court of Exchequer, and the accounts were kept, much as Robinson Crusoe kept his calendar on the desert island. In the course of considerable revolutions of time... still official routine inclined to these notched sticks, as if they were pillars of the constitution, and still the Exchequer accounts continued to be kept on certain splints of elm wood called 'tallies'. In the reign of George III. an inquiry was made by some revolutionary spirit, whether pens, ink, and paper, slates and pencils, being in existence, this obstinate adherence to an obsolete custom ought to be continued, and whether a change ought not to be effected.
>
> All the red tape in the country grew redder at the bare mention of this bold and original conception, and it took till 1826 to get these sticks abolished. In 1834 it was found that there was a considerable accumulation of them; and the question then arose, what was to be done with such worn-out, worm-eaten, rotten old bits of wood? I dare say there was a vast amount of minuting, memoranduming, and despatch-boxing, on this mighty subject. The sticks were housed at Westminster, and it would naturally occur to any intelligent person that nothing could be easier than to allow them to be carried away for fire-wood by the miserable people who live in that neighbourhood. However, they never had been useful, and official routine required that they never should be, and so the order went forth that they were to be privately and confidentially burnt. It came to pass that they were burnt in a stove in the House of Lords. The stove, overgorged with these preposterous sticks, set fire to the panelling; the panelling set fire to the House of Lords; the House of Lords set fire to the House of Commons; the two houses were reduced to ashes; architects were called in to build others; we are now in the second million of the cost thereof; the national pig is not nearly over the stile yet; and the little old woman, Britannia, hasn't got home to-night.[50]

Dickens remained unimpressed with politicians throughout his life. 'My faith in the people governing,' he said in 1869, 'is, on the whole infinitesimal; my faith in the people governed is, on the whole, illimitable.'[51]

The Devil's Acre only disappeared with the Slum Clearance Acts of the 1870s. Victoria Street—the main thoroughfare between Westminster and Pimlico—now stands on part of the site. The Peabody Housing Trust's Old Pye Street and Abbey Orchard Street development cleared all the filth away in the direction of Millbank from 1878, and still exists today, providing social housing in the heart of Westminster, in its handsome red and black striped, and yellow and grey brick, buildings.[52]

The fate of the records of the House of Commons and the Augmentation Office did not pass unnoticed. Within a couple of days, the loss of records was realized to be much more extensive than initially thought.[53] Following the disaster, Henry Cole repackaged and labelled all the Ministers' accounts in the Augmentation Office—some 4,000 or 5,000 parchment rolls—and catalogued them so they were in a better state of arrangement than before the fire. The boxes he used would protect them in the case of another disaster during which they might have to be thrown out of the windows again. It would take only 30 minutes to evacuate.[54] The 1836 inquiry into the Record Commission arose directly out of the fire and the agitation of Henry Cole. 'We must be obliged to bear the burthen of such a charge', wrote Edward Protheroe, the Commissioner who had been present on the night of the fire, the charge being that that Commission was indifferent to the risk to the nation's historic records, as were Parliament and the public as a whole.[55] The result was that the Record Commission was abolished, on account of its incompetence and corruption. 'With respect to the risk of injury from fire,' declared a further inquiry, the 1837 Royal Commission on Public Records,

> recent experience must have taught all persons who have attended to the subject, and has particularly impressed on our minds, that the

greatest precaution ought to be used, especially when it is considered that, as most of these documents exist but in single copies, their loss is irreparable, and that with every one of them disappears the evidence of some fact in the transactions of former ages.[56]

This led directly to the Public Record Office Act of 1838, which placed the Master of the Rolls, a senior judge, in charge of all the public records.[57] That then resulted in the building of the Public Record Office in Chancery Lane in the 1850s, which consolidated and protected government records on a dedicated site, including those saved from the Augmentation Office, and those residing in the Chapter House of the Abbey. The first Keeper of Public Records was none other than Francis Palgrave, who had anxiously watched the fire from the Chapter House roof four years earlier. And James Braidwood advised on the new Record Office's fire protection, with its multiple iron-doored stone repository rooms, more like prison cells, shelved with fireproof slate rather than wood.[58] William Black, the archivist who had helped at the Augmentation Office evacuation, became an Assistant Keeper of the Public Record Office in 1840.[59] The Exchequer records now form one of the most important foundation collections of the National Archives at Kew.

Henry Cole himself became Senior Assistant Keeper of the Public Record Office in 1838. His energy and vision stretched far beyond the confines of Chancery Lane and he became one of the most influential civil servants of the nineteenth century, involved in creating the first penny-post system, and the first Christmas card; campaigning for separate freight and passenger trains and a uniform narrow gauge on the railways; on the executive committee of the 1851 Great Exhibition working directly for Prince Albert; chairman of the Society of Arts and deviser of curricula for schools of industrial art and design; driving force behind the development of South Kensington into London's museum quarter after the death of Prince Albert; and establisher of the Albert Hall.[60]

The Acts, Journals, and other records of the House of Lords which remained in the Jewel Tower were moved out in 1864, into the Victoria Tower of the new Palace, completed in 1860, and purpose-built as a repository for the archives of Parliament. In fact, the Victoria Tower was proposed as the original location of the Public Record Office, before the Chancery Lane location won favour. In the early twentieth century, the surviving pre-1834 records of the House of Commons—just 231 original Journals—were moved to the Tower. An engraving of the Death Warrant of Charles I was republished by Mr Nethercliffe the lithographer, shortly after its narrow escape.[61] The original remained on display in the new House of Lords Library from 1851 to 1947, at which point it was transferred to the custody of the newly created House of Lords Record Office, now the Parliamentary Archives.

The melted weights and measures were recast by the scientific instrument-maker William Simms, after exhaustive trials and testing, creating the most accurate set of standards ever known.[62] The recasting required by the fire damage also provided an opportunity to completely reform the United Kingdom's standard system, which up until then had been crippled by two parallel imperial systems—troy and avoirdupois. Following an inquiry from 1838 to 1841, avoirdupois won, and troy weights were thereafter tightly restricted to gold, silver, and precious stones. The 1841 report also recommended, for the first time, introducing a decimal currency for the country, and the decimalization of length and weight (it was proposed, for example, that one hundred pounds should in future be called a *centner*).[63] These radical ideas finally began to bear fruit some 130 years later. In 1866 legislation was passed which transferred responsibility for the weights and measures standards from the Exchequer to the Board of Trade. In 1869, the Board of Trade Standards Department took over the vacated Jewel Tower at Westminster to house the standards, its construction being ideal for the purpose of calibration and measurement, being 'free from vibration and not liable to sudden fluctuations in temperature'. The standards finally moved out in 1938.[64]

The Rickmans had to move out of their undamaged house at the end of August 1835. It was then knocked down, as were the old Exchequer buildings containing the former offices of the Exchequer of Receipt. The eastern, river, side of New Palace Yard was cleared that autumn from Westminster Bridge to Westminster Hall to provide space for a potential new Palace. In time, that site made way for the arcade in front of Barry's Clock Tower, today better known as 'Big Ben'. Mrs Rickman died in 1836, the same year that her daughter Frances got married. John Rickman died aged 69 in 1840, and is buried beside his wife in St Margaret's Church.[65] After Manners-Sutton failed to be re-elected by his peers in the new Parliament, the Speaker's House ceased to be used for official accommodation. It was used first as the venue for consideration of the competition entries for the new Palace of Westminster, and then as Commons committee rooms until its demolition in 1842 to make way for the new House of Commons.[66] The Painted Chamber was finally demolished in 1847, soon after the MPs moved into their new accommodation. The old Lords Chamber, the Lesser Hall, was pulled down in September 1851, followed shortly afterwards by the bulk of Soane's Royal Gallery, Library and Parliament Office.[67] The Cloisters were restored and ready for new roofs in 1851.[68] That same year, under the immense pressure imposed by the building of the new Palace, the Office of Woods and Forests ceased to be a semi-independent government agency, and was transformed into the plain Office of Works, fully and tightly controlled by the government.[69] The restoration of the under-chapel of St Stephen's was undertaken in 1858–9, in an almost-secret operation. Regarded by some as one of the best examples of thirteenth-century architecture in the country, its renewal progressed despite the determination of the House of Commons by that date to bear down on the huge costs already incurred by the building of the new Palace. Working on a wooden platform illuminated by gas-jets, workmen repaired the calcined stonework vaulting and bosses, and then the roof was redecorated by Barry's son Edward in an exuberant gothic styling, at a total cost of £2,200. Those who criticized the waste of public money were reminded that 'it was not economy but scandalous parsimony which grudged what was necessary to support our national monuments' and that they ought to 'witness the enjoyment that the crypt afforded to the thousands of working men that walked through it'.[70]

The fire did not damage the responsible and clever John Phipps's career. He oversaw the perilous work of shoring up and disposal of the ruins.[71] In 1845 his annual salary was raised from £500 to £650. He continued to overwork, managing operations in Windsor and Richmond Park as well as the central London properties for which he was responsible. In the 1840s he designed additional coach houses at the Royal Mews, Pimlico; outbuildings, stables, and the entrance lodges at Frogmore; a boathouse and storerooms at Windsor; and other additions to the Chelsea Barracks (then an orphanage for soldiers' children), and on the Rolls estate in Chancery Lane where the Public Record Office was soon to be based. His job brought him into contact with Charles Barry. On occasions, with an eye on the purse-strings, he was critical of Barry's expenditure on the new Palace and elsewhere, but he worked with him as well (for example, on improvements to Horse Guards). He died in 1868, aged about 72.[72]

In pursuit of the great national undertaking to create a new Palace of Westminster, the *Morning Chronicle* wrote in early November, 'the people would be far from grudging a sufficient sum of money for the construction of an edifice which, combining magnificence with convenience, would reflect honour on our architectural science, and be an object of admiration to foreigners, and become a lasting and imperishable memorial to this and future generations'.[73] The opportunity was unparalleled. Just as the Great Fire of 1666 had rid London of the Black Death, here was another devastating fire which had rid the city of 'plague of great annoyance: we shall at last have Houses fit for the dispatch of their great public business'.[74] There was no question of moving from politically sacred Westminster. The old, battered site had a 'peculiar sanctity', was 'hallowed in our hearts', and had 'a halo about its reminiscences'.[75] The fervent wish was that 'the new House of Commons, by duties justly estimated and well performed, deliver down to their posterity associations as noble, as those which, with all its faults, and follies, and weaknesses, and crimes, yet remain to hallow the recollection of the old'.[76]

Smirke had been ordered to draw up plans for a rebuilt Houses of Parliament, of 'a moderate and suitable scale of magnitude'. He presented them to the King in February 1835, who passed them on, with his approval, to the new Prime Minister, Peel. A tidal wave of complaint about this procedure then engulfed the government. Smirke was accused of turning a repair job into a brand new commission, and of a lack of vision. Peel was accused of favouring his long-term client and friend. The Office of Woods and Forests was accused of being a closed shop.[77] In the face of a press campaign, pamphleteering by Sir Edward Cust, and intense pressure from—as usual—Joseph Hume, the commission was instead turned into an open competition to be judged by intelligent amateurs and men of taste outside the usual government circles. The old methods, which smacked of 'jobbery' and mediocrity, gave way for the first time to an anonymized, public award for which anyone could compete, and which would be awarded on merit. Cust himself purchased the old panelling from the former Exchequer buildings—the infamous Star Chamber—in 1836 before their demolition and installed it in his home at Leasowe Castle in the Wirral. The building is now a hotel, and the panelling can still be seen on display there.

'Incombustibility should be sought after as much as possible in the new buildings', advised the newspapers, somewhat superfluously.[78] At the end of July 1835, the government announced the launch of a competition to find a design for a new Palace of Westminster. The closing date was 1 December. Ninety-seven entries were submitted. On 29 February 1836, the winning design, number 64, identified only by a portcullis symbol (later discovered to be Charles Barry, with drawings undertaken by A. W. N. Pugin—in a neat irony, a Catholic), was selected by the competition commissioners.[79] The process of forgetting the 1834 fire and the amazing old Palace it ruined had begun. Over the next three decades a new Palace was built, a gigantic project characterized not only by genius but also by delays, massive overspend, slander and accusation, political interference, madness, and death. But that is another story.

DRAMATIS PERSONAE

Adelaide (1792–1849)   Queen of the United Kingdom of Great Britain and Ireland, wife of William IV.

Aldin   a doorkeeper, House of Lords Library.

Althorp, Lord (1782–1845)   3rd Earl Spencer, Whig, Chancellor of the Exchequer.

Auckland, Lord (1784–1849)   Whig, First Lord of the Admiralty.

Bankes, Henry (1757–1834)   Tory MP for Corfe Castle before the Great Reform Act; critic of the Office of Works.

Bankes, William (1786–1855)   his son, also a Tory MP.

Barry, Charles (1795–1860)   a highly successful architect.

Beaumonth, Barber   Superintendent, County Fire Office.

Bellamy, John (*fl.* 1773)   founder of Bellamy's refreshment rooms in the Palace of Westminster.

Bellamy, John (*fl.* 1834)   his son, and proprietor of Bellamy's at the time of the fire.

Mrs Bellamy   Bellamy junior's wife.

Bellingham, John (1770–1812)   assassin of Spencer Perceval.

Billings, Robert (1813–74)   watercolourist who captured the ruined Palace in paint in the days after the fire.

Birch   a fireman.

Black, William (1808–72)   archivist, Record Commission.

Braidwood, James (1800–61)   Superintendent, London Fire Engine Establishment.

Brandish, Frances   official Housekeeper of the House of Lords, a sinecure post.

Bristow   a fire engine manufacturer.

Britton, John (1771–1857)   antiquary and author.

Bromer, David   responsible for helping to obtain the Floating Engine.

Brougham, Lord (1778–1868)   Whig, Lord Chancellor, leading light in the passage of the Great Reform Act.

Broughton, Lord (1786–1869)   First Commissioner of Woods and Forests, the government office in charge of royal palaces, among other things.

Burch, Edward   fire engine keeper, Globe Insurance Co.

Campbell, Lord (1779–1861)   Whig, Attorney-General.

Carlyle, Thomas (1795–1881)   historian and critic.

Carter, John (1748–1817)   antiquarian and vehement critic of Wyatt's alterations to the House of Commons.

Chance (d.1835)   a celebrity canine. Parentage uncertain.

Chandler, William   a fireman.

Charles I   King of England and Scotland (1625–49), executed following trial in Westminster Hall.

Cobbett, William (1763–1835)   a Radical MP, journalist, and fierce critic of the accommodation arrangements in the Palace.

Cole, Henry (1808–82)   archivist, Record Commission.

Conboy, Thomas   Deputy Turncock, Chelsea Waterworks Co.

Constable, John (1776–1837)   landscape artist of genius.

Cooper, Charles Purton (1793–1873)   Secretary, Record Commission.

Cox   Clerk of the Furniture, Palace of Westminster.

Croker, John Wilson (1780–1857)   a Tory MP and fierce critic of the accommodation arrangements in the Palace; founder of the Athenaeum Club.

Cross, Joshua   plumber's mate employed by the Office of Woods and Forests, tally-burner.

Dickens, Charles (1812–70) a Parliamentary reporter and journalist. Aspiring novelist.

Dobson fire-engine keeper, St Martin's-in-the-Fields parish.

Duncannon, Lord (1781–1847) Whig, Home Secretary; First Commissioner of Woods and Forests to July 1834.

Edward I King of England (1272–1307), redecorated the Painted Chamber and began St Stephen's Chapel.

Edward III King of England (1327–77), completed work on St Stephen's Chapel, and endowed it as a secular college of canons in 1348.

Edward VI King of England (1547–53), dissolved the college at St Stephen's and offered use of its building to members of the House of Commons in 1548.

Ellis a servant of the Rickmans.

Evans, Robert fireman, Norwich Union fire office.

Fawkes, Guy (*c.*1570–1606) one of a band of discontented Catholic plotters who attempted to blow up Parliament and James I in 1605.

FitzClarence, Adolphus (1802–56) illegitimate son of William IV. Brother of the Earl of Munster and Frederick FitzClarence.

FitzClarence, Frederick (1799–1854) another illegitimate son of William IV, brother of the above.

Forty churchwarden, St Margaret's Church, Westminster.

Fremantle, Charles naval officer and brother of Thomas.

Fremantle, Thomas (1798–1890) Tory MP for Buckingham and brother of the above.

Furlong, Patrick paviour's mate, Irish, tally-burner.

George IV King of the United Kingdom of Great Britain and Ireland (1820–30).

Gregorie, David magistrate of the court at Queen's Square, Westminster.

Grey, Lord (1764–1845) Whig Prime Minister who pushed the Great Reform Act through Parliament.

Gurtkin   a friend or neighbour of the Rickmans.

Hambleton   a sub-engineer, LFEE.

Hannah   a servant or younger relative of the Rickmans.

Heath, William (1794/5–1840)   illustrator and caricaturist.

Henry III   King of England (1216–72), responsible for greatly expanding the Palace of Westminster, including the Painted Chamber, and rebuilding Westminster Abbey.

Henry VIII   King of England (1509–47), ceased to use the Palace of Westminster as a royal residence after a fire in 1512.

Herland, Hugh (d.1406?)   master carpenter responsible for the hammer-beam roof in Westminster Hall in the 1390s.

Haydon, Benjamin (1786–1846)   an ambitious artist.

Hill   a fireman.

Hill, Lord (1772–1842)   Commander-in-Chief of the army and epicure.

Hobhouse, John Cam, *see* **Broughton, Lord.**

Holroyd, William   a plumber, former employer and referee of Joshua Cross.

Hook, William   captain, Royal Navy, passing the Palace at the time of the fire.

Howard of Effingham, Lord (1536–1624)   Elizabethan naval commander, commissioned the Armada Tapestries.

Hubble, Joseph   Principal Turncock, Chelsea Waterworks Co.

Hume, Joseph (1777–1855)   Radical MP and fierce critic of the accommodation arrangements in the Palace.

Ireland, John (1761–1842)   Dean of Westminster Abbey.

Jane   a waitress, Bellamy's.

Jane   a servant or younger relative of the Rickmans.

Johnstone, Robert   contract paviour to the Office of Woods and Forests, working on repairs to Westminster Hall at the time of the fire. Boss of Patrick Furlong.

Jukes, John   foreman 'matter' in the Palace of Westminster.

Kirby   a labourer at the Palace of Westminster.

Knight, James   Deputy Turncock, Chelsea Waterworks Co.

Leary, John Frederick   Librarian, House of Lords.

Lee, Adam (c.1772–1843)   Clerk of Works at the Palace of Westminster from 1806 to 1832; aspiring architect.

Lefroy   David Bromer's solicitor.

Ley, John   Clerk of the House, House of Commons.

Lott   fire engine manufacturer.

Manby, George (1765–1854)   inventor of the first fire extinguisher, among other lifesaving innovations.

Manners-Sutton, Charles (1780–1845)   Speaker, House of Commons.

Manners-Sutton, Lady Ellen (?1791–1845)   his wife.

Manners-Sutton, Sir Charles   his son.

Manners-Sutton, Charlotte   his daughter.

Manners-Sutton, Henry (1814–1877)   another son.

Martin, Jonathan (1782–1838)   mentally ill arsonist who set fire to York Minster in 1829.

Mayne, Richard (1796–1868)   Magistrate (i.e. Commissioner) of the Metropolitan Police.

Melbourne, Lord (1779–1848)   Whig, Prime Minister.

Merryweather   fire engine manufacturer.

Milne, Alexander   Third Commissioner of the Board of Woods and Forests; formerly its Secretary. Boss of John Phipps.

Morgan, Bill   engineer of *The Fly*, towing the Floating Engine from Rotherhithe.

Moyes   a doorkeeper, House of Lords, and Receiver of peers' letters.

Mrs Moyes   his wife.

Mullencamp   another doorkeeper, House of Lords.

Mrs Mullencamp   his wife.

Munster, Earl of (1794–1842)   illegitimate eldest son of William IV. Brother of Adolphus and Frederick FitzClarence.

Murless, William   keeper of the St John's parish fire engine.

Nash, John (**1752–1835**)   architect and town planner, responsible for Regent's Park, the development of the West End, and Buckingham Palace.

Nicholas   the maitre d' at Bellamy's.

Owen, Richard   fire engine keeper, St Margaret's parish.

Palgrave, Francis (**1788–1861**)   Keeper of Public Records in the Chapter House, Westminster Abbey.

Palmer, James   labourer, Chelsea Waterworks Co.

Peel, Robert   Tory MP; Leader of the Opposition at the time of the fire.

Perceval, Spencer (**1762–1812**)   Tory Prime Minister assassinated in 1812.

Phillipps, Thomas (**1792–1872**)   obsessive manuscript and book collector.

Phipps, John (**d.1872**)   Assistant Surveyor, Office of Woods and Forests. In charge of the management of London royal palaces. Boss of Richard Weobley.

Protheroe, Edward   a Record Commissioner and MP.

Pugin, Augustus Welby Northmore (**1812–52**)   up-and-coming designer and architectural draughtsman.

Pulman, James   Yeoman Usher of the Black Rod, House of Lords, responsible for the safety and security of the building.

Reynolds, David   odd job boy in the House of Lords; son of Richard Reynolds.

Reynolds, Richard   firelighter to the House of Lords.

Rice, Frank   master of *The Fly*, towing the Floating Engine from Rotherhithe.

Richard II   King of England (1377–99), commissioned the hammer-beam roof in Westminster Hall.

Rickman, Anne   daughter of John and Susannah Rickman.

Rickman, Frances   daughter of John and Susannah Rickman.

Rickman, John (**1771–1840**)   Clerk Assistant, House of Commons, formerly Speaker's Secretary.

Rickman, Susannah   his wife.

Rooke, James   foreman, County Fire Office.

Rowan, Charles (1782?–1852)   Magistrate (i.e. Commissioner) of the Metropolitan Police.

Saunders, Edward   a fireman, British Fire Office.

Shuter, John   a tourist visiting the Palace on 16 October 1834.

Smirke, Robert (1780–1867)   attached architect to the Office of Woods and Forests, overseeing repairs to Westminster Hall at the time of the fire.

Smirke, Sydney (1798–1877)   his brother.

Snell, John   a tourist visiting the Palace on 16 October 1834.

Soane, John (1778–1837),   architect, responsible for neoclassical redevelopments to the Palace in the 1820s, and the building of the Law Courts along the west wall of Westminster Hall.

Solomons   a LFEE fireman.

Stanfield, Clarkson (1793–1867)   landscape artist.

Stone Smith, Henry (1819–74)   Clerk of Committees, House of Lords, and saviour of the Lords' archives.

Swing, Captain   mythical leader of the agrarian unrest movement of the early 1830s.

Taylor, Henry (1800–86)   colonial office administrator, poet, and friend of the Rickmans.

Tree   a LFEE fireman.

Turner, Joseph Mallord William (1775–1851)   landscape artist of genius.

Vardon, Thomas (1799–1867)   Librarian, House of Commons.

Villiers, Edward   friend of Henry Taylor.

Wellington, Duke of (1769–1852)   Tory Prime Minister to November 1830, fierce opponent of the Great Reform Act.

Weobley, Richard   Clerk of Works, Palace of Westminster, responsible for its maintenance and day-to-day building works.

West   a LFEE fireman.

Westmacott, Richard (1775–1856)   sculptor.

White, Charles engineer, British Fire Office.

Wildes, the family of the Keeper of the Exchequer, resident in the Palace, friends of the Rickmans.

William II 'Rufus' King of England (1087–1100), commissioned the building of Westminster Hall.

William IV King of the United Kingdom of Great Britain and Ireland (1830–7).

Williams a LFEE fireman.

Wren, Christopher (1632–1723) architect and Surveyor of the King's Works. Radically altered St Stephen's Chapel for use by the House of Commons.

Wright, Elizabeth mother-in-law of Jane Wright, standing in for her on the day of the fire.

Wright, George probably the husband of Jane Julia Wright; also possibly the name of Elizabeth Wright's husband, a former Lords doorkeeper, the latter either retired or deceased.

Wright, Jane Julia Deputy Housekeeper to the House of Lords. Absent on the day of the fire.

Wyatt, James (1746–1813) architect and Surveyor of the King's Works. Made radical and controversial alterations inside and out to much of the surviving Palace from 1800 onwards.

Yevele, Henry (d.1400) master mason responsible for alterations to Westminster Hall in the 1390s.

NOTES

Prologue

1. *Gentleman's Magazine*, November 1834, p. 477.
2. *The Times*, 17 October 1834, p. 2.
3. Trinity College, Cambridge, Dawson Turner papers, O.14.13/105, fo. 2r.
4. *The Times*, 23 October 1834, p. 2: 'According to Lord Brougham, the fire was an appropriate close of the session of immortality; the two Houses having reached the acme of glory, may have committed *felo de se* by spontaneous combustion, lest the honour should be sullied hereafter.'
5. Philip Ziegler, *William IV* (London, 1971), 250; A. N. Wilson, *The Victorians* (London, 2002), 10. These comments may in fact be apocryphal as I have been unable to track down the original source of either.
6. *The Examiner*, 19 October 1834, p. 659.
7. Benjamin Ferrey, *Recollections of A. W. N. Pugin and his Father Augustus Pugin* (London, 1978), 248.

Chapter 1

1. H. M. Colvin et al., *The History of the King's Works*, 6 vols. (London, 1963–82), vi. 126, 185–6, 188, 206–7, 647, 676, 679 [hereafter *KW*]. See Dorian Gerhold, *Westminster Hall: Nine Hundred Years of History* (London, 1999), 30, 35 for the King's Champion.
2. *Report of the Lords of the Council respecting the Destruction by Fire of the Two Houses of Parliament: with the minutes of evidence*, House of Commons Sessional Papers, 1835 (1), 14; *KW* vi. 220 n. 3, 542, 680 [hereafter *Report of the ... Fire*].
3. *KW* vi. 666–7. Cost based on the share of GDP calculator at www.measuringworth.com.
4. Christopher Thomas, Robert Cowie, and Jane Sidell, *The Royal Palace, Abbey and Town of Westminster on Thorney Island: Archaeological Excavations (1991–8) for*

the London Underground Ltd Jubilee Line Extension Project, Museum of London
Archaeological Service Monograph 22 (London, 2006), 155; John Goodall,
'The Medieval Palace of Westminster', in C. Riding and J. Riding (eds),
The Houses of Parliament: History, Art, Architecture (London, 2000), 49–67.

5. *KW* i. 528–33; Goodall, 'The Medieval Palace', 51, 59–61; Gerhold, *Westminster Hall*, 1.
6. *KW* v. 388; Gerhold, *Westminster Hall, passim.*
7. *KW* vi. 502–3, 666; Gerhold, *Westminster Hall*, 62–3.
8. John Crook and Roland B. Harris, 'Reconstructing the Lesser Hall: An
Interim Report from the Medieval Palace of Westminster Research
Project', in Clyve Jones and Sean Kelsey (eds), *Housing Parliament*, Parliamentary History Yearbook Trust (Edinburgh, 2002), 23, 39–40; *KW* v. 391.
The Lesser Hall was also known as the White Hall; throughout this book
I have used Lesser Hall to avoid confusion.
9. Crook and Harris, 'Reconstructing the Lesser Hall', 29–30.
10. Paul Binski, *The Painted Chamber at Westminster*, Society of Antiquaries
Occasional Paper IX (London, 1986), 1.
11. *KW* i. 501; *KW* v. 391; Goodall, 'The Medieval Palace', 55.
12. *KW* vi. 520–1; Soane's initial 1794 plans of the old House of Lords and its
'cellar' are at PA, HL/PO/JO/10/7/980A.
13. *KW* i. 493, 510–27; Goodall, 'The Medieval Palace', 56–7, 62.
14. *KW* v. 404.
15. *KW* vi. 525–6.
16. *KW* i. 491–2.
17. TNA, WORK 29/20–7.
18. PA, RIC/2/2, fo. 2v; *KW* vi. 517–19.
19. *KW* v. 389–90.
20. *KW* vi. 504–11.
21. *Report of the…Fire*, 12.
22. *KW* vi. 189–90.
23. *Report of the…Fire*, 12.
24. *Report of the…Fire*, 12–13.
25. *Report of the…Fire*, 12–20, 25–6, 40–1, 50. Weobley's evidence to the later
inquiry is confused and inconsistent: he muddles times and days of meetings, and claims to have discussed the destruction with Milne himself, and
also that Phipps had told him directly to find some other place than the
yard when he objected to it. None of this appears to be true.
26. *Report of the…Fire*, 22, 25, 28, 42.

27. *Report of the…Fire*, 19, 25–6, 43–4.
28. *Report of the…Fire*, 16, 24, 51.
29. J. Grant, *Random Recollections of the House of Commons from the year 1830 to the close of 1835*, 5th edn. (London, 1837), 1–2.
30. V. E. Chancellor, 'Hume, Joseph (1777–1855)', *Oxford Dictionary of National Biography* (Oxford, 2004 [hereafter *ODNB*]). *Parliamentary Debates*, 3rd Series, xix, col. 60; Grant, *Recollections of the House of Commons*, 6–7.
31. Alasdair Hawkyard, 'From Painted Chamber to St Stephen's Chapel: The Meeting Places of the House of Commons at Westminster until 1603', in Clyve Jones and Sean Kelsey (eds), *Housing Parliament*, Parliamentary History Yearbook Trust (Edinburgh, 2002), 79.
32. *Report from the Select Committee of House of Commons buildings, together with the minutes of evidence*, House of Commons Sessional Papers, 1831 (308), 8–10.
33. Ibid.; Grant, *Recollections of the House of Commons*, 2–3.
34. Quoted in Maurice Hastings, *Parliament House: The Chambers of the House of Commons* (London, 1950), 115; Ian Dyck, 'Cobbett, William (1763–1835)', *ODNB*.
35. *Parliamentary Debates*, 3rd Series, xvi, cols 370–2.
36. *Report from the Select Committee on the House of Commons buildings with the minutes of evidence taken before them*, House of Commons Sessional Papers, 1833 (269).
37. Ibid. 59.
38. *Parliamentary Debates*, 3rd Series, xvi, cols 370–9; xix, cols 59–62.
39. Ibid., col. 64.
40. Ibid., cols 64–6.
41. *Report…on the House of Commons buildings* (1833), 6.
42. *Parliamentary Debates*, 3rd Series, xix, cols 62, 66.
43. *The Times*, 18 October 1834, p. 5.

Chapter 2

1. *Report of the…Fire*, 25, 39, 52.
2. *Report of the…Fire*, 42, 52.
3. Lynn Hollis Lees, *Exiles of Erin: Irish Migrants in Victorian London* (Manchester, 1979), 57.
4. Michael Brock, *The Great Reform Act* (London, 1973), 33, 51–5.
5. Ibid. 33, 51–61.
6. E. Hobsbawm and G. Rudé, *Captain Swing* (London, 1969), 98, 85, 215.
7. T. Balston, *The Life of Jonathan Martin, Incendiary of York Minster* (London, 1945), 110.

8. *The Times*, 27 October 1834, p. 2.
9. TNA, HO 44/27 fo. 213.
10. *Report of the…Fire*, 21; the furnaces can clearly be seen on TNA, WORK 29/21 and 22, and also in Adam Lee's ground plan of the Palace drawn in 1807: Museum of London, A15453.
11. Colin G. C. Tite, 'The Cotton Library in the Seventeenth Century and its Manuscript Records of the English Parliament', *Parliamentary History*, xiv (1995), 123.
12. *Report of the…Fire*, 37.
13. *Report of the…Fire*, 19, 25–6, 43–4.
14. *Report of the…Fire*, 20–1, 25–6. Cross is vague or obstructive (as usual) about the size.
15. *Report of the…Fire*, 26, 37, 52.
16. *Report of the…Fire*, 19, 40, 41.
17. *Report of the…Fire*, 19, 25–6, 42–4.
18. *Report of the…Fire*, 38–9.
19. Peter Cunningham, *Murray's Handbook for Modern London, or London as it is* (London, 1851), 148.
20. *Report of the…Fire*, 38.
21. Goodall, 'The Medieval Palace', 49; *KW* i. 498, 505, 527.
22. *KW* iv. 286–8.
23. *KW* iv. 288–9.
24. Hawkyard, 'From Painted Chamber to St Stephen's Chapel', 77–80.
25. Brock, *Great Reform Act*, 19–20. By comparison, the 2001–5 Parliament had 659; the 2005–10 Parliament, 645.
26. Ibid. 18.
27. Ibid. 22–3.
28. Chancellor, 'Hume, Joseph', *ODNB*.
29. Brock, *Great Reform Act*, 29–30.
30. Ibid. 35, 39–40.
31. *Parliamentary Debates*, 3rd Series, i, cols 52–3.
32. Stanley H. Palmer, *Police and Protest in England and Ireland 1780–1850* (Cambridge, 1988), 310.
33. *Parliamentary Debates*, 3rd Series, i, cols 558–9.
34. For a summary list of dates see Brock, *Great Reform Act*, 391–2.
35. Norman Gash, *Aristocracy and People: Britain 1815–1865* (London, 1979), 366.
36. Dorothy Thompson, *The Chartists: Popular Politics in the Industrial Revolution* (Aldershot, 1984), 13.

37. See Philip Salmon, 'The English Reform Legislation, 1831–1832', in D. R. Fisher (ed.), *The History of Parliament: The House of Commons 1820–1832*, 7 vols. (Cambridge, 2009), i. 374–412, for a summary of the very complex effects of the Act; also Philip Salmon, *Electoral Reform at Work: Local Politics and National Parties 1832–1841* (Woodbridge, 2002), 252–5.

38. Edward Pearce, *Reform! The Fight for the 1832 Reform Act* (London, 2003), 294, 296.

39. *Report of the . . . Fire*, 28. Mistranscribed by the inquiry reporter as 'New-Pitt Street' (no such street existed).

40. Palmer, *Police and Protest*, 291.

41. J. A. Yelling, *Slums and Slum Clearance in Victorian London* (London, 1986), 1.

42. Richard J. Schiefen, 'Wiseman, Nicholas Patrick Stephen (1802–1865)', *ODNB*.

43. *Flora Tristan's London Journal: A Survey of London Life in the 1830s. A Translation of Promenades dans Londres*, ed. Dennis Palmer and Giselle Pincetl (London, 1980), 5, 79.

44. Jerry White, *London in the 19th Century* (London, 2007), 453.

45. Ibid. 457.

46. *Report from the Select Committee on the House of Commons buildings with the minutes of evidence taken before them*, House of Commons Sessional Papers, 1833 (269), 71.

47. Yelling, *Slums*, 20.

48. *Flora Tristan's London Journal*, 66–9.

49. *KW* v. 385–6.

50. White, *London in the 19th Century*, 452.

51. Ibid. 406.

52. Ibid. 38.

53. Ibid. 85.

54. Ibid. 23–6.

55. Ibid. 26.

56. Marie Busco, 'Westmacott, Sir Richard (1775–1856)', *ODNB*.

57. R. Bernal, 'My Aunt Mansfield', *The Keepsake* (London, 1835), 1.

58. *Flora Tristan's London Journal*, 4; Bernal, 'My Aunt Mansfield', 1; Dana Arnold, 'Burton, Decimus (1800–1881)', *ODNB*.

59. *Parliamentary Debates*, 3rd Series, xix, col. 62; James Grant, *Random Recollections of the House of Lords from the year 1830 to the close of 1836* (London, 1836), 39–40.

60. Daniel Hahn, *The Tower Menagerie* (London, 2003), 235.

Chapter 3

1. *Report of the . . . Fire*, 26, 34–6, 44–5.
2. *Report of the . . . Fire*, 21.
3. *Report of the . . . Fire*, 34–6, 40. Richard Reynolds' evidence is some of the most telling (and apparently honest) given to the inquiry.
4. *Report of the . . . Fire*, 29.
5. J. C. Sainty, 'The Office of Housekeeper of the House of Lords', *Parliamentary History*, 27 (2008), 258–9. Sainty has missed the fact that Jane Wright's mother-in-law was in charge on 16 October 1834.
6. *Report of the . . . Fire*, 59.
7. *Report of the . . . Fire*, 29, 31.
8. *Report of the . . . Fire*, 30–1.
9. *Report of the . . . Fire*, 31.
10. *Report of the . . . Fire*, 45.
11. *Report of the . . . Fire*, 26–7.
12. *Report of the . . . Fire*, 27, 29.
13. Brock, *Great Reform Act*, 243–4.
14. *Report of the . . . Fire*, 27.
15. *Report of the . . . Fire*, 29.
16. *Report of the . . . Fire*, 29, 32
17. *Report of the . . . Fire*, 28–9.
18. Adrian Tinniswood, *The Polite Tourist: A History of Country House Visiting* (London, 1998), 34–8.
19. *Report from the Select Committee on the Losses of the Late Speaker and Officers of the House by Fire of The Houses of Parliament; with the minutes of evidence and appendix*, House of Commons Sessional Papers, 1837 (493), 20.
20. *Report of the . . . Fire*, 30.
21. O. C. Williams, *The Clerical Organisation of the House of Commons 1661–1851* (Oxford, 1954), 240–3.
22. *An Act to regulate the Salaries of the Officers of the House of Commons, and to Abolish the Sinecure Offices of Principal Committee Clerks and Clerks of Ingrossments*, 4 & 5 William IV, *c*. 70.
23. Williams, *Clerical Organisation*, 276–7.
24. *Report of the . . . Fire*, 13, 22, 39.
25. *KW* vi. 121.
26. *KW* vi. 118.
27. *KW* vi. 664–6; 175–9.

28. *KW* vi. 157–8.

29. *KW* vi. 179–86.

30. *KW* vi. 120, 679; *Report… on House of Commons buildings* (1833), 82–3.

31. *KW* vi. 189, 191.

32. *Report of the… Fire*, 19–20.

33. *Guide to the Contents of the Public Record Office* (London, 1963), i. 45–6; *Dialogus de Scaccario*, ed. C. Johnson (Oxford, 1983), 7.

34. John Hudson, 'Richard fitz Nigel (c.1130–1198)', *ODNB*; *Dialogus de Scaccario*.

35. Much of what follows is taken from the anonymous 'Exchequer of Receipt: Original Receipts: E 402', National Archives Catalogue Introductory Note (London, 1995), in fact written by the author. See also Hilary Jenkinson, 'Exchequer Tallies', *Archaeologia*, 62 (1911), 367–80, and M. T. Clanchy, *From Memory to Written Record: England 1066–1307*, 2nd edn (Oxford, 1993), 123–4, the latter also picking up on the symbolism of administrative 'reform' and the burning of Parliament as embodied by the tallies.

36. *Dialogus de Scaccario*, 7.

37. *An Act for establishing certain Regulations in the Receipt of His Majesty's Exchequer*, 1783, 23 George III, *c*. 82.

38. Ibid., s. 2.

39. J. C. Sainty, *Officers of the Exchequer*, List and Index Society, Special Series, vol. 18 (London, 1983), 4, 13, 19; *Guide to the Contents of the Public Record Office*, i. 96.

40. *An Act to regulate the office of the Receipt of His Majesty's Exchequer at Westminster*, 1834, 4 & 5 William IV, *c*. 15.

41. *KW* i. 540–2; iv. 293–4; for their position in 1834, see TNA, WORK 29/21.

42. *KW* vi. 497–9.

43. *The Times*, 31 October 1834, p. 4.

44. Gillian Darley, *John Soane: An Accidental Romantic* (London, 1999), 287–9; *KW* vi. 499, 504–12.

45. *KW* vi. 505.

46. *Report of the… Fire*, 38; *KW* vi. 351.

47. *Report from the Select Committee on Record Commission*, House of Commons Sessional Papers, 1836 (565), 427–8.

48. Darley, *John Soane*, 314; E. W. Brayley and J. Britton, *The History of the Ancient Palace and Late Houses of Parliament at Westminster* (London, 1836), 409–10.

Chapter 4

1. *Gentleman's Magazine*, November 1834, p. 477.
2. *Report of the…Fire*, 52–3.
3. *Report of the…Fire*, 30–1.
4. *Report of the…Fire*, 52.
5. *Report of the…Fire*, 53.
6. *Report of the…Fire*, 32.
7. *Report of the…Fire*, 28, 47.
8. *Report of the…Fire*, 30, 32, 53.
9. *Report of the…Fire*, 28.
10. *Report of the…Fire*, 20.
11. *Report of the…Fire*, 21, 24.
12. *Report of the…Fire*, 21.
13. *Report of the…Fire*, 22.
14. *Report of the…Fire*, 23.
15. *Report of the…Fire*, 28.
16. *Report of the…Fire*, 45.
17. *Report of the…Fire*, 30, 36. For the location of the Grand Staircase, see TNA, WORK 29/21 and the Lobby is marked 'Waiting Room' in WORK 29/23.
18. *Report of the…Fire*, 30.
19. *The Times*, 21 October 1834, p. 3.
20. *Report of the…Fire*, 30.
21. *House of Lords Journal*, vol. 63, p. 932 (22 August 1831).
22. *Report of the…Fire*, 23, 36–7.
23. *Report of the…Fire*, 23, 28, 36, 37.
24. *An Act for the better regulation of Chimney Sweepers and their Apprentices, and for the safer construction of Chimneys and Flues*, 4 & 5 William IV, *c.* 35: ss. 2, 8, 15, 17, 18.
25. Act 3 & 4 William IV, *c.* 49; Act 4 & 5 William IV, *c.* 67.
26. Act 3 & 4 William IV, *c.* 103; Act 4 & 5 William IV, *c.* 1.
27. Act 4 & 5 William IV, *c.* 76.
28. *Report of the…Fire*, 40–1; *Morning Chronicle*, 29 October 1834.
29. See Thomas Allen, *The History and Antiquities of London, Westminster and Southwark and parts adjacent*, 5 vols. (London, 1828), iv. 201. The Painted Chamber was used often for conferences of the two Houses, or committees, 'there being a gallery of communication for the members of the House of Commons to come up without being crowded'.

30. *Report of the…Fire*, 40–1.
31. David Hughson, *Walks through London including Westminster and the borough of Southwark*, 2 vols. (London, 1817), ii. 205.
32. Allen, *The History and Antiquities of London*, iv. 200–1.
33. *Report of the…Fire*, 40–1.
34. Stephen Farrell, 'The Armada Tapestries in the Old Palace of Westminster', *Parliamentary History*, 29 (2010), 417–18, 420.
35. *Report of the…Fire*, 40–1.
36. *Report of the…Fire*, 30, 32.
37. *Report of the…Fire*, 37.
38. Parliamentary Works of Art Collection, WOA 1331.
39. I am grateful to Norman Davis for information on how chimney fires progress, the likely heat of coal and wood fires, and the melting point of copper, and also to John Peen.
40. *Report of the…Fire*, 21.
41. *Report of the…Fire*, 40.
42. *Report of the…Fire*, 37; Pullman's initial statement is at TNA, PC 1/2515.
43. *Report of the…Fire*, 37–8, 52.
44. *Report of the…Fire*, 28.
45. *Report of the…Fire*, 19, 21.

Chapter 5

1. *Report of the…Fire*, 23; *View of the Destruction of both Houses of Parliament, by fire; and immense loss of national property from a drawing taken on the spot by an eminent artist* (London, published by J. Fairbairn, 110 Minories). This is a printed broadsheet with a vivid handcoloured woodcut: Palace of Westminster Art Collection, WOA 2978a; *Gentleman's Magazine*, November 1834, pp. 477, 481; Brayley and Britton, *Ancient Palace*, 409.
2. *Report of the…Fire*, 30.
3. *Report of the…Fire*, 30, 37; *The Times*, 28 October 1834, p. 6.
4. *Report of the…Fire*, 23–4; *KW* vi. 118. Guy Fawkes' staircase lay at the south-east corner of the Painted Chamber: see H. M. Colvin, 'Views of the Old Palace of Westminster', *Architectural History*, 9 (1966), 163. Weobley's evidence to the inquiry on the fire is a little confused as to the timing of events, given his panic. However, it is fairly clear that what he is describing on the exterior at first is an ordinary chimney fire about 6 p.m., which he goes to investigate, and discovers it is much more serious inside the building. Only after this, is there a second phase outside, about 6.30 p.m.—which is when,

according to eyewitness accounts in the newspapers, the chimney fire develops into something more serious and the building bursts into flame more generally.

5. *Report of the ... Fire*, 30.
6. *Report of the ... Fire*, 24.
7. Sally Holloway, *Courage High! A History of Firefighting in London* (London, 1992), 8, 19–22, 27, 33.
8. Ibid. 23; Act 6 Anne, *c.* 58.
9. Charles Dickens, 'Our Parish', *Sketches by Boz* (London, 1836), ch. 1.
10. *View of the Destruction*; *The Times*, 17 October 1834, p. 3.
11. *Gentleman's Magazine*, November 1834, p. 481.
12. *The Examiner*, 19 October 1834.
13. *Report of the ... Fire*, 30. Lead melts at around 500°C.
14. *View of the Destruction*.
15. London Metropolitan Archives, LME/01 no. 498.
16. *View of the Destruction*.
17. *Morning Chronicle*, 18 October 1834.
18. This was George Manby, viewing the flashover from across the river. Trinity College, Cambridge, Dawson Turner papers, O.14.13/112, fo. 1v.
19. *Metropolitan Magazine*, November 1834, p. 291.
20. I am grateful to Norman Davis and Paul Kierans, Fire Safety Officers at Parliament, for information about the likely cause of blue flames, flashover, and other phenomena mentioned in this chapter and elsewhere.
21. *The Times*, 18 October 1834, p. 5. The timing is corroborated by Charles Patten of 15 Medway Street, Horseferry Road, who wrote to Duncannon, the Home Secretary, saying that he saw the burning at its commencement at 6.30 p.m. (TNA, HO 44/27, fo. 211).
22. *View of the Destruction*.
23. *The Times*, 18 October 1834, p. 5; *Morning Chronicle*, 18 October 1834.
24. *View of the Destruction*.
25. TNA, WORK 11/26/5, fos. 4, 19–29.
26. Ibid., fos. 19–29.
27. Ibid., fos. 21, 23.
28. Ibid., fo. 28.
29. Ibid., fo. 29.
30. Ibid., fos. 23, 25–6.
31. Ibid., fos. 21, 22, 24.
32. Ibid., fos. 20, 27.

33. Ibid., fo. 27.

34. Ibid., fo. 19.

35. *Report of the…Fire*, 25. It seems likely that the 'government engines' referred to below by Captain Hook, RN when he arrived between 6.45 p.m. and 7 p.m. are the Board of Works ones from Horse Guards (see TNA, London, HO 44/29, fo. 51r). Interestingly there does appear to have been a House of Commons fire engine during the reign of George I (see Lucy Worsley, *Courtiers* (London, 2010), 157) but no sign of one in 1834. I am grateful to Mark Purcell for this reference.

36. *Report…on the Losses of the Late Speaker*, 9.

37. TNA, WORK 11/26/5, fos. 9v–10r.

38. *View of the Destruction.*

39. *The Times*, 17 October 1834, p. 3.

40. *Bristol Mercury*, 18 October 1834.

41. F. E. C. Gregory, 'Rowan, Sir Charles (1782?–1852)', *ODNB*; Clive Emsley, 'Mayne, Sir Richard (1796–1868)', *ODNB*.

42. Palmer, *Police and Protest*, 296–8, 303.

43. D. Ascoli, *The Queen's Peace: The Origins and Development of the Metropolitan Police* (London, 1979), 90, 97; Palmer, *Police and Protest*, 309.

44. *The Times*, 17 October 1834, p. 3. For the arrival of police and soldiers at this time, see also *Gentleman's Magazine*, November 1834, p. 481.

45. Palmer, *Police and Protest*, 309.

46. *Bristol Mercury*, 18 October 1834.

47. *The destruction of both Houses of Parliament by fire, October 16 1834 As seen from Abingdon Street*, Printed by W. Belch, Bridge Street, Borough (private collection).

48. TNA, London, HO 44/29, fo. 51r.

49. *Recollections of a long life by Lord Broughton (John Cam Hobhouse) with additional extracts from his private diaries*, ed. Lady Dorchester, 6 vols. (London, 1911), v (1834–40), 21.

50. PA, RIC/3, fo. 1v.

51. PA, RIC/1, RIC/3 (plan).

52. *Report of the…Fire*, 28, 46, 48.

53. *Report of the…Fire*, 6.

54. *Gentleman's Magazine*, November 1834, p. 481.

55. *Manchester Times & Gazette*, 17 October 1834.

56. W. M. Torrens, *Memoirs of William Lamb, Second Viscount Melbourne*, new rev. edn (London, 1890), 303–4.

57. Durham University Library, GRE/B41/2/145/2, 24 October 1834.

Chapter 6

1. London, Guildhall Library, MS 15728/2: London Fire Establishment Committee Minutes, 20 October 1834, p. 80.
2. For this and what follows, see W. F. Hickin, *Organised against Fire: A Short Organisational History of the London Fire Brigade and its Predecessors from 1833 to 1996* (n.p., 1996), 3–4; James Braidwood, *Fire Prevention and Fire Extinction* (London, 1866), 15; M. C. Curtoys, 'Braidwood, James (1800–1861)', *ODNB*.
3. Holloway, *Courage High!*, 42–4.
4. *The Times*, 18 February 1835, p. 6.
5. London, Guildhall Library, MS 15728/2, pp. 80–1.
6. *The Times*, 17 October 1834, p. 3; TNA, WORK 29/21, 23.
7. *View of the Destruction*.
8. *The Times*, 17 October 1834, p. 3.
9. *Morning Chronicle*, 17 October 1834.
10. *The Examiner*, 19 October 1834; *Morning Chronicle*, 17 October 1834.
11. *View of the Destruction*.
12. Ibid.
13. *The Times*, 17 October 1834, p. 3.
14. *The Times*, 18 October 1834, p. 5.
15. *View of the Destruction*; *The Times*, 17 October 1834, p. 3; *Bristol Mercury*, 18 October 1834.
16. *Bristol Mercury*, 18 October 1834.
17. Ibid.
18. *View of the Destruction*; *The Times*, 17 October 1834, p. 3.
19. *The Examiner*, 19 October 1834.
20. *The Morning Chronicle*, 18 October 1834.
21. As in, for example, Herefordshire Record Office, G2/IV/J/66, 16 October 1834.
22. *The Times*, 17 October 1834, p. 3; *Gentleman's Magazine*, November 1834, p. 482.
23. *Morning Chronicle*, 17 October 1834; *Bristol Mercury*, 18 October 1834. Brief details of Gregorie's career are given at London Metropolitan Archives, MJP/R/082/1–2.
24. *View of the Destruction*.
25. *Report of the...Fire*, 20.
26. *Morning Chronicle*, 17 October 1834; *Bristol Mercury*, 18 October 1834; *The Examiner*, 19 October 1834.

27. Farrell, 'Armada Tapestries', 418–20, 435.
28. Ibid. 424, 431; Malcolm Hay, 'The Armada Paintings in the New Palace of Westminster', *Parliamentary History*, 29 (2010), 441–2. Farrell details a conjectural hanging for the tapestries in the old House of Lords, in the process revealing that not all of them could have been accommodated there.
29. *The Times*, 21 October 1834, p. 3.
30. Farrell, 'Armada Tapestries', 435–8.
31. Ibid. 427, 429; Roy Strong, *Lost Treasures of Britain* (London, 1990), 211–13.
32. *Report…on the Losses of the Late Speaker*, 18; W. R. McKay, *Clerks in the House of Commons 1363–1989: A Biographical List*, House of Lords Record Office Occasional Publications No. 3 (HMSO, London, 1989), 28, 39, 54, 120. Gunnell had died in February 1831.
33. *Report…on the Losses of the Late Speaker*, 19; McKay, *Clerks*, 81.
34. *Report…on the Losses of the Late Speaker*, 1, 20. Collins is not listed in McKay, *Clerks*.
35. *Report…on the Losses of the Late Speaker*, 18–19; McKay, *Clerks*, 65, 96. The *Gentleman's Magazine*, November 1834, p. 483, also mentions this, giving Walmisley the name of Dyson.
36. *Report…on the Losses of the Late Speaker*, 20.
37. PA, HL/PO/LB/2/1.
38. David Eastwood, 'Rickman, John (1771–1840)', *ODNB*.
39. PA, RIC/3, fo. 1v.
40. PA, HL/PO/LB/2/1.
41. *The Times*, 17 October 1834, p. 3. The rooms destroyed can be reconstructed from TNA, WORK 29/23. London, Guildhall Library, MS 15728/2, p. 81.
42. *Bristol Mercury*, 18 October 1834.
43. Rosemary Hill, *God's Architect: Pugin and the Building of Romantic Britain* (London, 2008), 127.
44. *The Collected Letters of A. W. N. Pugin*, ed. Margaret Belcher, 2 vols. (Oxford, 2001), i (1830–42), 42.
45. M. H. Port, 'Barry, Sir Charles (1795–1860)', *ODNB*; Alfred Barry, *Memoir of the Life and Works of the Late Sir Charles Barry, Architect*, 2nd edn (London, 1870), 145–6.
46. See, for example, Hill, *God's Architect*, 128, 546 n. 53.
47. London, Guildhall Library, MS 15728/2, pp. 80–1.

48. London Metropolitan Archives, LME/01 no. 498.
49. Hickin, *Organised against Fire*, 4.
50. London, Guildhall Library, MS 15728/2, p. 81.
51. *The Times*, 18 October 1834, p. 5.
52. TNA, WORK 29/21, 23.
53. *The Examiner*, 19 October 1834.
54. Ibid.; *The Times*, 18 October 1834, p. 5.
55. *The Examiner*, 19 October 1834.
56. London, Guildhall Library, MS 15728/2, p. 81.
57. Ibid.
58. *The Times*, 17 October 1834, p. 3. The original states Parliament Street instead of St Margaret's Street but this is an obvious error.
59. *Morning Chronicle*, 17 October 1834.
60. Christopher F. Lindsey, 'Braithwaite, John, the younger (1797–1870)', *ODNB*.
61. Holloway, *Courage High!*, 47.
62. *The Times*, 22 November 1834, p. 6. This writer mentions 12 LFEE engines, but we know from Braidwood's report that he used 13.
63. Trinity College, Cambridge, Dawson Turner papers, O.14.13/105, fo. 1r.
64. R. B. Prosser, rev. R. C. Cox, 'Manby, George William (1765–1854)', *ODNB*.
65. Trinity College, Cambridge, Dawson Turner papers, O.14.13/105, fo. 1v.
66. *The Times*, 17 October 1834, p. 3.
67. London, Guildhall Library, MS 15728/2, p. 81.

Chapter 7

1. *Gentleman's Magazine*, November 1834, p. 482; *Bristol Mercury*, 18 October 1834.
2. *The Collected Letters of A. W. N. Pugin*, i. 42.
3. *Gentleman's Magazine*, November 1834, p. 482.
4. *The Times*, 17 October 1834, p. 3; 18 October 1834, p. 5.
5. *Gentleman's Magazine*, November 1834, p. 482; *The Times*, 17 October 1834, p. 3.
6. *Bristol Mercury*, 18 October 1834.
7. *Fatal destruction of both Houses of Parliament, Thursday evening, October 16th 1834* (British Library, C.116.i.2, fo. 29).
8. Ibid.
9. *View of the Destruction*.

10. Grant, *Recollections of the House of Commons*, 2–3.
11. London Metropolitan Archives, LME/01 no. 498. Hickin, *Organised against Fire*, 4. A second floating engine was acquired in 1837 and this 'Upper Float' was moored just above Southwark Bridge. London, Guildhall Library, MS 15728/2, p. 81; *The Times*, 18 February 1835, p. 6.
12. *The Times*, 17 October 1834, p. 3.
13. Ibid. This report mistakenly mentions the three 'south' windows at the gable end of the chapel, forgetting that at Westminster, the Thames flows northwards not eastwards.
14. *View of the Destruction*.
15. *Gentleman's Magazine*, November 1834, p. 482.
16. Allen, *The History and Antiquities of London*, iv. 203.
17. *Bristol Mercury*, 18 October 1834.
18. Arthur Irwin Dasent, *The Speakers of the House of Commons* (London, 1911), 312.
19. *The Times*, 17 October 1834, p. 3.
20. *Gentleman's Magazine*, November 1834, p. 482. Howard's coffee-house is by mistake referred to as Hayes' coffee-house in this description.
21. *The Times*, 17 October 1834, p. 3.
22. *The Times*, 18 October 1834, p. 5; *View of the Destruction*; *The Examiner*, 19 October 1834.
23. TNA, WORK 11/26/5, fos. 7v–8r.
24. Ibid., fos. 9v–10r.
25. PA, HC/SA/SJ/9/52 and 55.
26. *View of the Destruction*.
27. *Gentleman's Magazine*, November 1834, p. 481.
28. *View of the Destruction*; *The Times*, 18 October 1834, p. 5.
29. *The Times*, 17 October 1834, p. 3.
30. Anne Sebba, *The Exiled Collector: William Bankes and the Making of an English Country House* (Wimborne Minster, 2009), 149–56.
31. Ibid. 148, 157, 162–70, 174–8.
32. *View of the Destruction*.
33. Norwich, Norfolk Record Office, MC 81/26/222 525x8.
34. *The Times*, 17 October 1834, p. 3.
35. PA, HL/PO/RO/1/158, pp. 2–3; Thomas Seccombe, 'Wornum, Ralph Nicholson (1812–1877)', rev. David Carter, *ODNB*.
36. *The Times*, 17 October 1834, p. 3; *View of the Destruction*.
37. *Bristol Mercury*, 18 October 1834.

38. *Morning Post*, 17 October 1834.

39. White, *London in the 19th Century*, 265.

40. Simon Werrett, *Fireworks: Pyrotechnic Arts and Sciences in European History* (Chicago and London, 2010), 242.

41. P. H. Marshall, *Charles Marshall RA, his Origins, Life and Career* (privately printed, Truro, 2000), 25–6.

42. Adam Lee, *Description of the Cosmoramic and Dioramic Delineations of the Ancient Palace of Westminster and St Stephen's Chapel* (London, 1831): copy preserved at PA, LGC/5/3. See also Mireille Galinou, 'Adam Lee's Drawings of St Stephen's Chapel, Westminster. Antiquarianism and Showmanship in Early 19th Century London', *Transactions of the London and Middlesex Archaeological Society*, 34 (1983), 231–44.

43. White, *London in the 19th Century*, 265.

44. *View of the Destruction*.

45. *Gentleman's Magazine*, November 1834, p. 478.

46. PA, HL/PO/RO/1/78; Glennis Byron, 'Landon, Letitia Elizabeth (1802–1838)', *ODNB*.

47. *Morning Chronicle*, 18 October 1834; *Morning Post*, 18 October 1834.

48. *The Times*, 17 October 1834, p. 3. The Scots Fusilier Guards were also used as auxiliary firemen, and their commander Sir John Aitchison received Duncannon's thanks for their work: see J. Lunt, 'Aitchison, Sir John (1789–1875)', *ODNB*.

49. *Life of Benjamin Robert Haydon, Historical Painter, from his Autobiography and Journals*, ed. Tom Taylor, 3 vols. (London, 1853), ii. 355, 362–3; Robert Woof, 'Haydon, Benjamin Robert (1786–1846)', *ODNB*.

50. Katherine Solender, *Dreadful Fire! Burning of the Houses of Parliament* (Cleveland Museum of Art, Cleveland, Ohio, 1984), 11.

51. *View of the Destruction*.

52. Solender, *Dreadful Fire!*, 42; *Morning Chronicle*, 18 October 1834.

53. *Morning Post*, 18 October 1834.

54. Solender, *Dreadful Fire!*, 42. Extract from the shorthand diary of John Green Waller, FSA.

55. Solender, *Dreadful Fire!*, 42; Pieter van der Merwe, 'Stanfield, Clarkson (1793–1867)', *ODNB*.

56. Wilson, *The Victorians*, 13.

57. London, Tate Britain, Turner Bequest, Finberg CCLXXXIV and CCLXXXIII.

58. Solender, *Dreadful Fire!*, 42–3, 57; Strong, *Lost Treasures of Britain*, 213.

59. C. R. Leslie, *Memoirs of the Life of John Constable*, ed. J. Mayne, 3rd edn (London, 1995), 203, 302.

60. Trinity College, Cambridge, Dawson Turner papers, O.14.13/105, fo. iv.

61. Solender, *Dreadful Fire!*, 60.

62. *Gentleman's Magazine*, November 1834, p. 478; *The Times*, 17 October 1834, p. 3; *View of the Destruction*; *Metropolitan Magazine*, November 1834, p. 291.

63. *Metropolitan Magazine*, November 1834, pp. 291, 293.

64. Curtoys, 'Braidwood, James (1800–1861)', *ODNB*.

65. *Morning News*, 16 October 1835; *Preston Chronicle*, 19 December 1835. His collar couplet may also have read (depending on which story you believe): 'Stop me not, but let me jog | I'm the Fire Establishment dog'.

66. Parliamentary Works of Art Collection, WOA 589, 1667 and 1990; two larger portraits of Chance were created by William Heath, and published by Rudolph Ackermann: see *New Sporting Magazine*, January 1835, p. 204.

67. Dickens, *Sketches by Boz*, ch. 18. For an example of this phenomenon see the tale of an MP said to have been in committee room 21 over the House of Commons Library, who allegedly first discovered the fire at 6 p.m., gave the alarm, and saved all the nearby books and papers: *The Times*, 18 October 1834, p. 5. Another tells of the discovery of the fire by Mr Cottle in the Bishop's Lobby: *Morning Chronicle*, 18 October 1834.

68. *The Times*, 21 October 1834, p. 3, 25 October 1834, p. 2; 30 October 1834, p. 3.

69. *The Letters of Charles Dickens*, i (1820–1839), ed. Madeline House and Graham Storey (Oxford, 1965), 40–3.

70. Michael Slater, 'Dickens, Charles John Huffnam (1812–1870)', *ODNB*.

71. See *Dickens' Journalism*, ed. Michael Slater (London, 1994–2000), ii. 373. I am grateful to William Hale for this reference.

72. Charles Dickens, *Oliver Twist*, ch. 48.

Chapter 8

1. *View of the Destruction*.

2. *Gentleman's Magazine*, November 1834, p. 478.

3. *The Times*, 18 October 1834, p. 5; *Morning Chronicle*, 18 October 1834.

4. *Report... of the Fire*, 36.

5. *View of the Destruction*.

6. *Report... on the Losses of the Late Speaker*, 19.

7. Robert Brown, 'Bellamy, John (*fl.* 1773)', *ODNB* (online edn., 2005); J. P. W. Ehrman and Anthony Smith, 'Pitt, William (1759–1806)', *ODNB*; 'The House of Commons refreshment department', House of Commons

Information Office, Factsheet G19, General Series (2003), 2; O. C. Williams, *The Officials of the House of Commons* (London, 1909), 13.

8. Williams, *Officials*, 55.

9. TNA, WORK 29/23, 25–7.

10. Charles Dickens, 'A Parliamentary Sketch', *Sketches by Boz*, ch. 18.

11. *Report ... on the Losses of the Late Speaker*, 10, 19.

12. Charles Dickens, 'A Parliamentary Sketch', *Sketches by Boz*, ch. 18.

13. O. C. Williams, *Lamb's Friend the Census Taker: Life and Letters of John Rickman* (London, 1912), 310; Mark Reger, 'Taylor, Sir Henry (1800–1886)', *ODNB*.

14. PA, RIC/2/1, fo. 1v.

15. Williams, *Life and Letters of John Rickman*, 310.

16. PA, RIC/2/1, RIC/2/2, RIC/3. For the mattresses, see Chapter 13.

17. PA, RIC/2/1, fo. 1v.

18. *Gentleman's Magazine*, November 1834, p. 478.

19. *Ipswich Journal*, 18 October 1834.

20. *Metropolitan Magazine*, November 1834, p. 291.

21. Trinity College, Cambridge, Dawson Turner papers, O.14.13/105, fo. 1r.

22. *Gentleman's Magazine*, November 1834, p. 478; *View of the Destruction*.

23. *Metropolitan Magazine*, November 1834, pp. 291, 293.

24. Williams, *Life and Letters of John Rickman*, 312.

25. *Recollections of a long life by Lord Broughton*, v. 22.

26. *The Times*, 17 October 1834, p. 2.

27. *Manchester Times & Gazette*, 18 October 1834.

28. *Bristol Mercury*, 18 October 1834.

29. *The Times*, 17 October 1834, p. 3.

30. Brock, *Great Reform Act*, 69.

31. M. L. Bush, *The Casualties of Peterloo* (Lancaster, 2005), 2–3; D. Read, *Peterloo: The Massacre and its Background* (Manchester, 1958), 133.

32. Brock, *Great Reform Act*, 247–8, 251–3.

33. *Bristol Mercury*, 18 October 1834.

34. *Metropolitan Magazine*, November 1834, p. 291.

35. *Morning Chronicle*, 18 October 1834.

36. *Morning Post*, 18 November 1834.

37. *The Times*, 18 October 1834, p. 5: *The Standard*, quoted in *Cobbett's Weekly Register*, 25 October 1834, pp. 201–2.

38. Thomas Carlyle to Alexander Carlyle, 24 October 1834, *Letters of Thomas Carlyle 1826–1836*, ed. Charles Eliot Norton (London, 1889), 455. I am grateful to Philip Salmon for this reference.

39. Fred Kaplan, 'Carlyle, Thomas (1795–1881)', *ODNB*.
40. *Morning Chronicle* (Police Intelligencer), 18 October 1834.
41. *The Times*, 18 October 1834, p. 5.
42. And other papers picked it up, including the *Ipswich Journal*, 18 October 1834.
43. *Cobbett's Weekly Register*, 25 October 1834, pp. 199–200.
44. 17 October 1834. Quoted in *Cobbett's Weekly Register*, 25 October 1834, pp. 201–2. My italics. *The Examiner*, 17 October 1834, p. 659 repeated identical sentiments.
45. *Ipswich Journal*, 18 October 1834, quoting *The Standard*.
46. *Sir George Crewe, Squire of Calke Abbey: extracts from the journals of Sir George Crewe, 1815–1834*, ed. C. Colin Kitching (Matlock, 1995), 133. I am grateful to Philip Salmon for this reference.
47. *The Examiner*, 26 October 1834.
48. *Manchester Times & Gazette*, 18 October 1834.
49. British Library (formerly Fox-Talbot Museum, Lacock), Fox-Talbot papers, LA(H)34–012, 27 October 1834.
50. *Bristol Mercury*, 18 October 1834.
51. *The Times*, 18 October 1834, p. 5.

Chapter 9

1. *Metropolitan Magazine*, November 1834, p. 293.
2. *View of the Destruction*.
3. *Gentleman's Magazine*, November 1834, 477–8.
4. *Bristol Mercury*, 18 October 1834.
5. *View of the Destruction*.
6. *The Examiner*, 19 October 1834.
7. Aylesbury, Centre for Buckinghamshire Studies, D/FR/119/13/1/9. I am grateful to Philip Salmon for this reference. The brothers were Admiral Sir Charles Howe Fremantle, GCB (1800–69), and Thomas Francis Fremantle, first Baron Cottesloe (1798–1890): see *ODNB*.
8. Letter to *The Times* from 'C.H', 18 October 1834.
9. *Morning Chronicle*, 18 October 1834.
10. For this and what follows, see James Braidwood, *On the construction of fire-engines and apparatus, the training of firemen and the method of proceeding in cases of fire* (Edinburgh, 1830), 1, 3, 10; James Braidwood, *Fire Prevention and Fire Extinction* (London, 1866), 124–30, 133–4, 143, 144, 146.
11. Braidwood, *Fire Prevention and Fire Extinction*, 72–3.

12. *The Times*, 17 October 1834, p. 3; London, Guildhall Library, MS 15728/2, p. 81; Letter to *The Times* from 'C.H', 18 October 1834; *The Examiner*, 19 October 1834.

13. *Gentleman's Magazine*, November 1834, pp. 477–8.

14. *The Examiner*, 19 October 1834.

15. Parliamentary Works of Art Collection, WOA 1669.

16. *The Times*, 17 October 1834, p. 3; London, Guildhall Library, MS 15728/2, p. 81; Letter to *The Times* from 'C.H', 18 October 1834; *The Examiner*, 19 October 1834.

17. Braidwood, *Fire Prevention and Fire Extinction*, 17, 75.

18. Ibid. 73, 76, 77.

19. *Morning Chronicle*, 18 October 1834.

20. Letter to *The Times* from 'C.H', 18 October 1834.

21. *The Times*, 17 October 1834, p. 3.

22. Parliamentary Works of Art Collection, WOA 1645.

23. *Morning Chronicle*, 18 October 1834; Letter to *The Times* from 'C.H', 18 October 1834.

24. Braidwood, *Fire Prevention and Fire Extinction*, 79.

25. Letter to *The Times* from Braidwood, 18 October 1834.

26. *Life of John, Lord Campbell*, edited by his daughter Mrs Hardcastle, 2 vols. (London, 1881), 54, 17 October 1834.

27. Braidwood, *Fire Prevention and Fire Extinction*, 28, 78.

28. *Morning Chronicle*, 18 October 1834.

29. *The Times*, 17 October 1834, p. 3.

30. *The Examiner*, 19 October 1834.

31. Ibid. The blocked windows belonged to the Speaker's servants' quarters on the other side of the wall; thanks to Mark Collins for this information.

32. TNA, HO 44/29 fo. 51v.

33. *Morning Post*, 18 November 1834.

34. K. D. Reynolds, 'FitzClarence, George Augustus Frederick, first earl of Munster (1794–1842)', *ODNB*.

35. TNA, HO 44/29 fo. 51v.

36. *The Times*, 17 October 1834, p. 3; TNA, HO 44/29 fo. 51v.

37. *The Times*, 17 October 1834, p. 3.

38. TNA, HO 44/29, fo. 52r.

39. Ibid.

40. Ibid., fo. 52v.

41. Braidwood, *Fire Prevention and Fire Extinction*, 67.

42. *The Examiner*, 19 October 1834.

43. Gordon L. Teffeteller, 'Hill, Rowland, first Viscount Hill (1772–1842), *ODNB* (2004–8).

44. *The Times*, 22 October 1834, p. 3; *Gentleman's Magazine*, November 1834, p. 482.

45. Sydney Smirke, 'Remarks on the Architectural History of Westminster Hall', *Archaeologia*, 26 (1836), 406.

46. *The Times*, 24 October 1834, p. 3.

47. *The Examiner*, 19 October 1834.

48. *The Times*, 17 October 1834, p. 3.

49. Ibid.

50. *Morning Chronicle*, 17 October 1834.

51. *Gentleman's Magazine*, November 1834, p. 478.

52. *The Examiner*, 19 October 1834; *Gentleman's Magazine*, November 1834, p. 478.

53. Letter to *The Times* from 'C.H', 18 October 1834.

54. *The Examiner*, 19 October 1834.

55. *Life of John, Lord Campbell*, 55–6.

56. *The Examiner*, 19 October 1834.

57. *Morning Chronicle*, 18 October 1834; *The Times*, 18 October 1834, p. 5. For the exact location of committee room 12, see TNA, WORK 29/23.

58. *View of the Destruction.*

59. Magnificently detailed plans of the roof and the repairs made are at PA, HC/LB/1/114/42; Caroline Shenton et al., *Victoria Tower Treasures from the Parliamentary Archives* (London, 2010), 132.

60. This description is taken from Robin Demaus, 'Precision Treatment of Death Watch Beetle Attack' <http://www.buildingconservation.com/articles/beetle/beetle.html> (accessed January 2011).

61. Malcolm Hay and Rachel Kennedy, *The Medieval Kings c1388. An Exhibition of Three Statues from the South Wall. Their History and Recent Conservation* (London, 1993), 3, 15.

62. *The Collected Letters of A. W. N. Pugin*, 42.

63. *Morning Chronicle*, 18 October 1834.

Chapter 10

1. *The Times*, 17 October 1834, p. 3.

2. White, *London in the 19th Century*, 418.

3. *Morning Chronicle*, 18 October 1834.

4. London, Guildhall Library, MS 15728/2, p. 83; *Morning Chronicle*, 18 October 1834.

5. *The Times*, 18 October 1834, p. 5; *Morning Chronicle*, 18 October 1834. There is some confusion in the papers between Hambleton and Hamilton. They are in fact two separate firemen. Hambleton may be the 'Hamilton' wrongly reported to be killed when some flooring fell in at 11.30 p.m.; *View of the Destruction*.

6. *The Times*, 18 October 1834, p. 5.

7. *The Times*, 21 October 1834, p. 3; compensation claims are in TNA, WORK 11/26/5.

8. Trinity College, Cambridge, Dawson Turner papers, O.14.13/105, fo. iv.

9. *Morning Post*, 18 October 1834.

10. *Bristol Mercury*, 18 October 1834.

11. *Metropolitan Magazine*, November 1834, p. 294.

12. Phillipps's own letter in the third person, to *The Spectator*, 23 October 1834. Quoted in D. Menhennet, *The House of Commons Library: A History* (London, 1991), 17; A. N. L. Munby, *Phillipps Studies*, 2 vols. (London, 1971), i (III), 101–2.

13. *View of the Destruction*.

14. *Gentleman's Magazine*, November 1834, p. 478.

15. *View of the Destruction*.

16. *The Times*, 18 October 1834, p. 5; *Morning Chronicle*, 18 October 1834.

17. *Report...on the Losses of the Late Speaker*, 9; Norman Gash, 'Sutton, Charles Manners-, first Viscount Canterbury (1780–1845)', *ODNB*; PA, RIC/1.

18. PA, RIC/3, fo. iv.

19. *KW* vi. 532–3.

20. Allen, *The History and Antiquities of London*, iv. 204.

21. PA, OOW/3, fos. 56–63.

22. *Manchester Times & Gazette*, 18 October 1834; *The Times*, 17 October 1834, p. 3; *Morning Chronicle*, 18 October 1834.

23. *View of the Destruction*; *Report...on the Losses of the Late Speaker*, 9.

24. *Life of John, Lord Campbell*, 54; *View of the Destruction*.

25. *Morning Post*, 18 October 1834.

26. Ibid.

27. *Report...on the Losses of the Late Speaker*, 8.

28. *KW* vi. 534.

29. *Report...on the Losses of the Late Speaker*, 9.

30. *View of the Destruction*; *Report...on the Losses of the Late Speaker*, 9 inaccurately states that the dining furniture was consumed, but in fact it is listed as

surviving in PA, OOW/3, fo. 56. The post-fire picture of the crypt is Parliamentary Works of Art Collection, WOA 5195.

31. *The Examiner*, 19 October 1834.

32. J. T. Smith, *Antiquities of Westminster* (London, 1807), 264–6. I am grateful to Emma Gormley for this reference.

33. *Report of the...Fire*, 46.

34. *Morning Post*, 18 October 1834.

35. *Gentleman's Magazine*, November 1834, p. 483.

36. PA, HC/SO/1/6; G. F. R. Barker, 'Sutton, John Henry Thomas Manners-, third Viscount Canterbury (1814–1877)', rev. H. C. G. Matthew, *ODNB*.

37. PA, HC/SO/1/6. The Speaker and his wife were subsequently offered temporary residence in the house of Mr Bailey, MP, in Seymour Place: PA, RIC/3, fo. 1r.

38. PA, HC/SO/1/6; *Morning Chronicle*, 17 October 1834; *Morning Post*, 18 October 1834.

39. *Report...on the Losses of the Late Speaker*, 2; the Speaker himself said he arrived 12 or 13 hours after the fire commenced, the *Morning Post*, 18 October 1834, said he left Brighton at seven in the morning.

40. *The Holland House Diaries, 1831–1840*, ed. Abraham D. Kreigel (London, 1977), 266.

41. *Report...on the Losses of the Late Speaker*, 3, 9.

42. Ibid. 3.

43. *Metropolitan Magazine*, November 1834, p. 291.

44. *The Times*, 17 October 1834, p. 3.

45. *Morning Chronicle* (Police Intelligencer), 18 October 1834; *Morning Chronicle*, 25 October 1834; *Morning Post*, 18 October 1834.

46. PA, RIC/3, fo. 1v.

47. *Morning Chronicle*, 25 October 1834.

48. *View of the Destruction*; *Morning Chronicle*, 18 October 1834.

49. *The Times*, 22 October 1834, p. 3.

50. Ibid.

51. *The Times*, 22 November 1834, p. 6.

52. Ibid.

53. *The Morning Chronicle* (Police Intelligencer), 18 October 1834.

54. PA, RIC/2/1, RIC/2/2, RIC/3.

55. *Report...on the Losses of the Late Speaker*, 17; PA, RIC/2/2, fo. 1v.

56. Menhennet, *The House of Commons Library*, 1–2, 6–7; *KW* vi. 528–30.

57. *The Times*, 17 October 1834, p. 3.

58. *The Times*, 18 October 1834, p. 5; C. W. Earle, *Memoirs and Memories: Pot-pourri from a Surrey Garden* (London, 1911), 73.
59. *Morning Chronicle*, 17 October 1834.
60. *The Times*, 17 October 1834, p. 3; *View of the Destruction*.
61. Parliamentary Works of Art Collection, WOA 1978.
62. *The Times*, 17 October 1834, p. 3.
63. Menhennet, *The House of Commons Library*, 1.
64. *View of the Destruction*.
65. Menhennet, *The House of Commons Library*, 16–19; *Gentleman's Magazine*, November 1834, p. 483.
66. *View of the Destruction*; C. C. Pond, 'Vardon, Thomas (1799–1867)', *ODNB*.
67. Menhennet, *The House of Commons Library*, 16–19.
68. Rowan Strong, 'Campbell, Archibald (c.1669–1744)', *ODNB*.
69. *The Times*, 17 October 1834, p. 3.

Chapter 11

1. J. D. Cantwell, *The Public Record Office, 1838–1958* (London, 1991), 33.
2. *Guide to the Contents of the Public Record Office*, i. 81–2.
3. *Morning Post*, 17 October 1834, 18 October 1834.
4. *Metropolitan Magazine*, November 1834, p. 292.
5. Menhennet, *The House of Commons Library*, 17; Munby, *Phillipps Studies*, i (III), 101–2.
6. Alan Bell, 'Phillipps, Sir Thomas, baronet (1792–1872)', *ODNB*; Munby, *Phillipps Studies*, 140–1.
7. *Report from the Select Committee on Record Commission*, House of Commons Sessional Papers, 1836 (429), p. vi; Peter Walne, 'The Record Commissions, 1800–1837', *Journal of the Society of Archivists* (1960), 8–16.
8. J. A. Hamilton, 'Cooper, Charles Purton (1793–1873)', rev. Beth F. Wood, *ODNB*.
9. Cantwell, *Public Record Office*, 1–4.
10. *Report...on Record Commission*, 89, 99–100, 436–7; Elizabeth Bonython and Anthony Burton, *The Great Exhibitor: The Life and Work of Henry Cole* (London, 2003), 42; Bernard Nurse, 'Black, William Henry (1808–1872)', *ODNB*.
11. *Report...on Record Commission*, 33, 436–7.
12. Ibid. 33.
13. Ann Cooper, 'Cole, Sir Henry (1808–1882)', *ODNB*.
14. Cantwell, *Public Record Office*, 4.
15. See Chapter 9, n. 45.

16. *The Times*, 24 October 1834, p. 3; H. C. G. Matthew, 'Warburton, Henry (1784–1858)', *ODNB*. The Record Commissioner was named in the account as Charles Cooper, but this was before 10 p.m., and he did not arrive on the scene until after midnight. For Hume, see *View of the Destruction*.
17. *The Times*, 24 October 1834, p. 3.
18. Ibid.
19. E. M. Clerke, 'Somerville, William (1771–1860)', rev. Roger T. Stearn, *ODNB*.
20. *The Times*, 24 October 1834, p. 3; Somerville later claimed £5 costs for doing this work: see TNA, WORK 11/26/5, fos. 9v–10r.
21. *Bristol Mercury*, 18 October 1834.
22. *Reports from the Select Committee appointed to inquire into the state of the public records of the Kingdom etc* (London, 1800), 66–7.
23. Ibid.
24. PA, HL/PO/LB/2/1.
25. Williams, *Life and Letters of John Rickman*, 311–12.
26. Ibid. 131; PA, RIC/2/2, fos. 1v–2r.
27. *Morning Chronicle* (Police Intelligencer), 18 October 1834; *Morning Post*, 18 October 1834.
28. Trinity College, Cambridge, Dawson Turner papers, O.14.13/112.
29. Now London, British Library, Egerton MS 1048. Arnold Hunt of the British Library first suggested that the petition's water damage was related to the 1834 fire, when it went on display at the *Taking Liberties* exhibition in 2008.
30. Menhennet, *The House of Commons Library*, 16.
31. *Gentleman's Magazine*, November 1834, p. 483.
32. *Morning Post*, 17 October 1834.
33. *New Monthly Magazine*, vol. xlii, November 1834, p. 353.
34. Elizabeth Hallam Smith, 'The Chapter House as a Record Office', in Warwick Rodwell and Richard Mortimer (eds), *Westminster Abbey Chapter House: The History, Art and Architecture of 'a chapter house beyond compare'* (London, 2010), 125, 128–32.
35. *Metropolitan Magazine*, November 1834, p. 292.
36. *The Times*, 17 October 1834, p. 3.
37. A. P. Stanley, *Historical Memorials of Westminster Abbey* (London, 1911), 380–2; Tony Trowles, 'Ireland, John (1761–1842)', *ODNB*; G. H. Martin, 'Palgrave, Sir Francis (1788–1861)', *ODNB*.
38. Quoted in Hallam Smith, 'The Chapter House as a Record Office', 134.

39. Centre for Kentish Studies, U840 C534/2.

40. *The Times*, 21 October 1834, p. 3; *Gentleman's Magazine*, November 1834, p. 483.

41. *Recollections of a long life by Lord Broughton*, v. 21–2.

42. John Greenhead, 'History of the House of Lords Library', *House of Lords Library Note LLN 2009/005* (London, 2009), 1; *KW* vi. 524.

43. *View of the Destruction*; *The Times*, 24 October 1834, p. 3.

44. *View of the Destruction*.

45. *KW* vi. 525; *Morning Post*, 17 and 18 October 1834 for the suggestion that a builder's plumber in the Library had left a fire alight there.

46. D. L. Jones, 'The House of Lords and the Chambre des Pairs: An Exchange of Books, 1834', in H. S. Cobb (ed.), *Parliamentary History, Libraries and Records: Essays Presented to Maurice Bond* (House of Lords Record Office, London, 1981), 52.

47. TNA, HO 44/27, fo. 313, 15 November 1834.

48. A. J. Taylor, *The Jewel Tower, Westminster* (London, 1965), 5–6.

49. D. J. Johnson, 'The House of Lords and its Records, 1660–1864', in Cobb (ed.), *Parliamentary History, Libraries and Records*, 28–9. The petitions, orders, etc. are today known as the House of Lords Main Papers and are one of the most important series of the Parliamentary Archives (HL/PO/JO/10).

50. *The Times*, 21 October 1834, p. 3; *Morning Chronicle*, 17 October 1834; *Bristol Mercury*, 18 October 1834.

51. Johnson, 'The House of Lords and its Records', 29; PA, HL/PO/1/166; J. C. Sainty, *The Parliament Office in the Nineteenth and Early Twentieth Centuries: Biographical Notes on Clerks in the House of Lords 1800 to 1939* (House of Lords Record Office, London, 1990), 51; *First Report of the Royal Commission on Historical Manuscripts*, Command Paper C.55 (London, 1874), Appendix, pp. 1–3. The Commission's report on the location of the records at the time of the fire is rather confused and vague.

52. TNA, WORK 11/26/5, fos. 8v–9r. The claim was disallowed.

53. *Morning Post*, 18 October 1834; *The Examiner*, 19 October 1834. This account in fact reports that the records were stored under the Painted Chamber, mistaking it for the Royal Gallery. It seems unlikely, given the damage to the Painted Chamber.

54. PA, HL/PO/1/1.

55. H. S. Cobb, 'The Victoria Tower and the Records of Parliament, 1864–1986', *The Table: The Journal of the Society of Clerks at the Table in Commonwealth Parliaments*, liv (1986), 54–63.

56. *Fatal destruction of both Houses of Parliament* (British Library, C.116.i.2, fo. 29).

57. *The Standard,* quoted *Cobbett's Weekly Register,* 25 October 1834, pp. 201–2. A similar story of a sweep crying his trade at the top of his voice is given in the *Morning Post,* 18 October 1834.
58. *An Act for the better regulation of Chimney Sweepers and their Apprentices,* 4 & 5 William IV, *c.* 35, s. 15; *The Times,* 18 October 1834, p. 5.
59. *Bristol Mercury,* 18 October 1834.
60. *Ipswich Journal,* 18 October 1834.

Chapter 12

1. *Bristol Mercury,* 18 October 1834.
2. *The Times,* 18 October 1834, p. 5; *Morning Chronicle,* 18 October 1834.
3. *Bristol Mercury,* 18 October 1834.
4. *Morning Post,* 17 October 1834.
5. *Morning Chronicle,* 18 October 1834.
6. TNA, HO 44/27, fo. 305.
7. Ibid., fo. 315.
8. *Morning Chronicle,* 29 October 1834.
9. PA, HC/SO/1/6.
10. Mike Matthews, *Captain Swing in Sussex and Kent: Rural Rebellion in 1830* (Hastings, 2006), 2; Michael Holland (ed.), *Swing Unmasked* (Milton Keynes, 2005), 1, 2, 7, 9.
11. *Fatal destruction of both Houses of Parliament* (British Library, C.116.i.2, fo. 29).
12. *The Times,* 17 October 1834, p. 3; *Morning Post,* 17 October 1834.
13. Balston, *The Life of Jonathan Martin,* 61–3. For Martin's violent childhood and youth, his mental illness and religious fanaticism, see H. C. G. Matthew, 'Martin, Jonathan (1782–1838)', *ODNB.*
14. Balston, *The Life of Jonathan Martin,* 77, 85, 90, 94–5, 97. In fact, the so-called arson attempt at Westminster Abbey was an accidental fire caused by thieves trying to steal lead off the roof: see *Gentleman's Magazine,* April 1829, p. 363; *Children's Daily Telegraph* (25 April 1931). I am grateful to Christine Reynolds, Deputy Keeper of Westminster Abbey Muniments, for the latter two references.
15. *Morning Chronicle,* 18 October 1834.
16. *View of the Destruction.*
17. *Morning Chronicle,* 18 October 1834.
18. *The Times,* 17 October 1834, p. 3; *The Times,* 18 October 1834, p. 5. Guildhall Library, London, MS 15728/2, p. 8; TNA, WORK 11/26/5, fo. 22.

19. PA, HL/PO/LB/2/1.

20. TNA, WORK 11/26/5, fos. 9v–10r.

21. Braidwood, *Fire Prevention and Fire Extinction*, 20.

22. London Metropolitan Archives, London: LME/01 no. 498.

23. TNA, WORK 11/26/5 fos. 3, 5v–6, 17v, 30–1, 32, 38–49.

24. *The Collected Letters of A. W. N. Pugin*, i. 42.

25. *Gentleman's Magazine*, November 1834, p. 483.

26. *Morning Post*, 18 November 1834.

27. *The Times*, 17 October 1834, p. 3; *View of the Destruction*.

28. Michael Liversidge and Paul Binski, 'Two Ceiling Fragments from the Painted Chamber at Westminster Palace', *Burlington Magazine*, 137 (1995), 491–501.

29. Palace of Westminster Art Collection, WOA 1648.

30. *KW* vi. 513–14.

31. Liversidge and Binski, 'Two Ceiling Fragments', 491; *KW* vi. 519; Phillis M. Rogers, 'Medieval Fragments from the Old Palace of Westminster in the Sir John Soane Museum', in Cobb (ed.), *Parliamentary History, Libraries and Records*, 2.

32. *KW* i. 494–8; Paul Binski, *The Painted Chamber at Westminster*, Society of Antiquaries Occasional Paper IX (London, 1986), 6 and *passim*.

33. *KW* i. 499; Binski, *Painted Chamber*, 3–4, 115; Palace of Westminster Art Collection, WOA 922.

34. W. M. Ormrod, *The Reign of Edward III* (New Haven and London, 1990), 13–15; Hawkyard, 'From Painted Chamber to St Stephen's Chapel', 62–5, 67–9. 84.

35. *View of the Destruction*; *KW* vi. 520, 523–4.

36. PA, HL/PO/LB/2/1.

37. *Bristol Mercury*, 18 October 1834.

38. PA, RIC/2/1, fo. 1r.

39. *Bristol Mercury*, 18 October 1834.

40. Bedfordshire and Luton Archives and Record Service, Z629/41, 17 November 1834.

41. *Gentleman's Magazine*, November 1834, p. 483.

42. David C. Hamilton, *The Assassination of the Prime Minister: John Bellingham and the Murder of Spencer Perceval* (Stroud, 2008), 3, 84–5; P. J. Jupp, 'Perceval, Spencer (1762–1812)', *ODNB*; Michael J. Turner, 'Bellingham, John (1770–1812)', *ODNB*.

43. Chancellor, 'Hume, Joseph', *ODNB*.

44. *The Times*, 22 October 1834, p. 3.

45. PA, OOW/3, fo. 60.

46. Hastings, *Parliament House*, 141. Hastings is wrong to say only a few pieces of furniture survived the fire.

47. PA, OOW/3, fo. 61.

48. As filmed, for example, in the BBC Parliament documentary 'Burning down the House', October 2009.

49. *The Times*, 17 October 1834, p. 3.

50. Quoted in Williams, *Life and Letters of John Rickman*, 310.

51. PA, RIC/2/1, fo. 2r; HL/PO/LB/2/1.

52. PA, HL/PO/LB/2/1.

53. PA, RIC/3, fo. IV.

Chapter 13

1. PA, RIC/1. This letter was partly reproduced in Williams, *Life and Letters of John Rickman*, 308–12. Still owned by Mrs Lefroy (née Anne Rickman) in 1912, the Parliamentary Archives acquired the correspondence at auction in 1997.

Chapter 14

1. *Gentleman's Magazine*, December 1834, p. 629.

2. *An Act for ascertaining and establishing Uniformity of Weights and Measures* (1824), 5 George IV, *c*. 74.

3. Ibid., *s. 4; Minutes of evidence taken before the select committee on the bill to amend and render more effectual two acts of the fifth and sixth years of the reign of his late Majesty King George the Fourth, relating to Weights and Measures*, House of Commons Sessional Papers, 1834 (464), 1.

4. *Report of the Commissioners appointed to consider the steps to be taken for restoration of the standards of weight & measure*, House of Commons Sessional Papers, 1842 (356), 5–6.

5. *The Examiner*, 19 October 1834.

6. *The Times*, 17 October 1834, p. 3.

7. Guildhall Library, London, MS 15728/2, p. 82.

8. Braidwood, *Fire Prevention and Fire Extinction*, 33.

9. Ibid. 42–3.

10. Ibid. 61.

11. London Metropolitan Archives, LME/01 no. 498.

12. *The Times*, 18 October 1834, p. 5.

13. Guildhall Library, London, MS 15728/2, p. 80; *The Times*, 18 October 1834, p. 5.

14. Braidwood, *On the construction of fire-engines*, 13–14.

15. TNA, WORK 11/26/5, fos. 8v–9r.

16. Ibid., fos. 5v–6r.

17. Ibid., fos. 10v–11r.

18. *The Times*, 22 October 1834, p. 3.

19. Guildhall Library, London, MS 15728/2, pp. 82–3.

20. Aylesbury, Centre for Buckinghamshire Studies, D/FR/119/13/1/9: I am grateful to Philip Salmon for this reference.

21. TNA, WORK 11/26/5, fos. 5v–6. Cost of beer from www.oldbaileyonline. org/static/Coinage.jsp (accessed August 2010).

22. TNA, WORK 11/26/5, fos. 9v–10r.

23. *The Times*, 18 October 1834, p. 5. The original version of this report confuses the Commons Library with the Lords Library.

24. *The Times*, 18 October 1834, p. 5.

25. Parliamentary Works of Art Collection, WOA 2389.

26. Williams, *Life and Letters of John Rickman*, 311–12.

27. PA, RIC/2/2, fo. 2r.

28. TNA, WORK 29/766, 769, 770. These are the preliminary drawings for Frederick MacKenzie's 1844 publication on the Antiquities of St Stephen's.

29. *KW* vi. 601.

30. TNA, WORK 11/26/5, fos. 9v–10r.

31. See Chapter 10. TNA, WORK 11/26/5, fos. 9v–10r.

32. TNA, WORK 11/26/5, fos. 9v–10r.

33. Ibid., fos. 10v–11r.

34. Ibid., fos. 9v–10r.

35. See Chapter 10. TNA, WORK 11/26/5, fos. 9v–10r.

36. See average earnings calculator at www.measuringworth.com.

37. TNA, WORK 11/26/5, fos. 5v–11r.

38. Ibid., fos. 10v–11r; *The Times*, 21 October 1834, p. 3.

39. TNA, WORK 11/26/5, fos. 10v–11r.

40. Ibid., fos. 8v–9r, 13–14v; *Report...on the Losses of the Late Speaker*, 10, 19.

41. TNA, WORK 11/26/5, fo. 15.

42. Ibid., fos. 10v–11r, 16.

43. Ibid., fos. 6v–7r, 16.

44. Ibid., fos. 10v–11r.

45. Ibid., fo. 5.

46. *Report ... on the Losses of the Late Speaker*, 20.

47. TNA, WORK 11/26/5, fo. 17v.

48. 'Oh What a Flare Up', *The Rambler's Flash Songster, nothing but out and outers, adapted for gentlemen only, etc.* (London, [1838?]), 32–4. My thanks to Judith Flanders for drawing this to my attention.

49. 'The Burning of the Houses of Lords and Commons', quoted in Solender, *Dreadful Fire!*, 11.

50. P. Cormack, 'The Great Fire of Westminster 1834', *The Historian* (Autumn, 1984), 3–4.

51. University of Chicago Library, Illinois, USA, Song and Broadsheet Collection, box 1, fo. 9.

Chapter 15

1. *The Times*, 18 October 1834, p. 5; *Morning Chronicle*, 18 October 1834.

2. *The Times*, 18 October 1834, p. 5; *Morning Chronicle*, 18 October 1834.

3. *Morning Chronicle*, 18 October 1834.

4. *Morning Post*, 18 November 1834.

5. *Morning Chronicle*, 18 October 1834; *The Times*, 18 October 1834, p. 5.

6. *Morning Chronicle*, 18 October 1834.

7. H. Maxwell (ed.), *The Creevey Papers: A Selection from the Correspondence and Diaries of the Late Thomas Creevy, MP* (London, 1903), ii. 288.

8. Menhennet, *The House of Commons Library*, 18.

9. Christopher Jones, *The Great Palace* (London, 1983), 67.

10. *The Times*, 18 October 1834, p. 5.

11. PA, OOW/3, fos. 35–49.

12. *Report ... on the Losses of the Late Speaker*, 2.

13. *The Times*, 18 October 1834, p. 5.

14. *Morning Post*, 18 November 1834.

15. PA, RIC/2/1, fo. 2r.

16. PA, HL/PO/LB/2/1.

17. PA, RIC/3, fo. 1v.

18. *The Times*, 18 October 1834, p. 5.

19. *The Times*, 21 October 1834, p. 3.

20. *The Examiner*, 19 October 1834.

21. *The Times*, 21 October 1834, p. 3.

22. *The Examiner*, 26 October 1834.

23. *Recollections of a long life by Lord Broughton*, v. 22.

24. *The Examiner*, 26 October 1834.
25. *Recollections of a long life by Lord Broughton*, v. 22; *Gentleman's Magazine*, November 1834, p. 482 (the royal visit is wrongly dated by the *Magazine* as occurring on 17 October).
26. PA, RIC/3, fo. 1r.
27. *The Examiner*, 26 October 1834.
28. *Recollections of a long life by Lord Broughton*, v. 22–3; *Gentleman's Magazine*, November 1834, p. 482.
29. *The Times*, 20 October 1834, p. 3; the *Morning Chronicle* and the *Morning Post*, 18 October 1834, warned their readers in advance.
30. PA, RIC/3, fo. 1v.
31. *The Times*, 20 October 1834, p. 3.
32. *The Times*, 21 October 1834, p. 3.
33. PA, RIC/2/1, fo. 2r.
34. PA, RIC/2/2, fo. 1r.
35. Guildhall Library, London, MS 15728/2, p. 80. Firemen's wages for the week ending 18 October had cost £101 2*s*. 6*d*.
36. PA, RIC/2/2, fo. 2.
37. *Morning Post*, 18 November 1834.
38. TNA, PC 1/2515.
39. *The Times*, 18 October 1834, p. 5; *Morning Chronicle*, 18 October 1834.
40. *Morning Chronicle*, 18 October 1834.
41. *Morning Post*, 18 November 1834.
42. *New Monthly Magazine*, November 1834, p. 352.
43. TNA, PC 1/2515.
44. *Bristol Mercury*, 18 October 1834.
45. *The Times*, 17 October 1834, p. 3.
46. Ibid.
47. *The Times*, 18 October 1834, p. 5.
48. *The Examiner*, 19 October 1834.
49. *The Examiner*, 26 October 1834.
50. *The Greville Memoirs 1814–1860*, ed. L. Strachey and R. Fulford, 8 vols. (London, 1939), iii. 89, 13 November 1834.
51. *The Times*, 21 October 1834, p. 3.
52. *Gentleman's Magazine*, November 1834, p. 482 (which wrongly dates the start of the inquiry as 22 October); *The Examiner*, 26 October 1834. For the garbled reports, see *The Times*, *Morning Chronicle*, and *The Examiner* itself, among others.

53. *Life of John, Lord Campbell*, 55–6.
54. *Morning Post*, 18 November 1834.
55. *Life of John, Lord Campbell*, 55–6; see a similar view of the inquiry from Lord Farnborough: Carlisle, Cumbria Record Office, D/LONS/L1/2/50.
56. *Morning Chronicle*, 24 October 1834.
57. *Report of the…Fire*, 43.
58. *Report of the…Fire*, 10.
59. See the business card in TNA, PC 1/2515.
60. *Report of the…Fire*, 2, 6, 10.
61. *Fatal destruction of both Houses of Parliament* (British Library, C.116.i.2. fo. 29).
62. *A Portion of the Journal Kept by Thomas Raikes Esq, from 1831 to 1847*, 4 vols. (London, 1856), i. 291–3.
63. *The Times*, 24 October 1834, p. 3; *Journal of the House of Lords*, lxvi (London, 1834), 997–8.
64. PA, RIC/2/2, fos. 1v, 2v.
65. *The Times*, 24 October 1834, p. 3. In this account Rickman is said to stand in for the Speaker—an error for Clerk of the House.
66. PA, HL/PO/1/1. Wright received £27 (£1 10s. per week) for his work as temporary gatekeeper between 3 November 1834 and 21 March 1835.
67. *Morning Chronicle*, 24 October 1834.
68. Williams, *Life and Letters of John Rickman*, 312.

Epilogue

1. *Morning Chronicle*, 24 October 1834. Buckingham Palace is described as 'the King's Palace' in the report.
2. *Morning Chronicle*, 24 October 1834.
3. *The Times*, 3 November 1834, p. 3.
4. TNA, WORK 11/26/5, fos. 10v–11r.
5. Cambridge, Trinity College, O.14.13/112.
6. TNA, HO 44/27, fos. 318–19.
7. *The Examiner*, 19 October 1834.
8. Ibid.
9. Guildhall Library, London, MS 15728/2, p. 83.
10. Curtoys, 'Braidwood, James (1800–1861)', *ODNB*.
11. *The Examiner*, 19 October 1834.
12. *Gentleman's Magazine*, December 1834, p. 629.
13. Annette Peach, 'Billings, Robert William (1813–1874)', *ODNB*.
14. PA, RIC/2/2, fo. 2v.

15. P. Jackson, 'Scharf, Sir George (1820–1895)', *ODNB*; Parliamentary Works of Art Collection, WOA 3793.

16. *The Mirror of Literature, Amusement and Instruction*, 22 November 1834, p. 368.

17. *Morning Chronicle*, 27 October 1834.

18. Cormack, 'Great Fire of Westminster 1834', 6.

19. *Gentleman's Magazine*, December 1834, pp. 629–30.

20. Ibid.

21. Tinniswood, *The Polite Tourist*, 136.

22. L. H. Cust, rev. Ralph Hyde, 'Marshall, Charles (1806–1890)', *ODNB*.

23. *The Times*, 4 November 1834, p. 3.

24. PA, LGC/5/3, fos. 341–4, 363.

25. *The Times*, 5 December 1834, p. 4.

26. PA, LGC/5/3, fo. 349. Another view was that the only value lay in the portraits, 'which perhaps may be considered inestimable', but this was a minority view (fo. 351).

27. PA, LGC/5/3, fo. 361; Farrell, 'Armada Tapestries', 440.

28. Hay, 'Armada Paintings', 441–51.

29. Solender, *Dreadful Fire!*, 21, 42, 61–2.

30. *Morning News*, 16 October 1835; *Preston Chronicle*, 19 December 1835.

31. *The Times*, 18 October 1834, p. 5; *The Times*, 19 October 1834; *Gentleman's Magazine*, November 1834, p. 480; *The Examiner*, 19 October 1834; *Morning Chronicle*, 18 October 1834; *Morning Post*, 18 November 1834.

32. *Gentleman's Magazine*, November 1834, p. 483.

33. *Morning Chronicle*, 18 October 1834.

34. *Caledonian Mercury*, 20 October 1834.

35. *The Times*, 24 October 1834, p. 3.

36. *Morning Chronicle*, 24 October 1834.

37. *Morning Chronicle*, 8 November 1834.

38. *The Holland House Diaries, 1831–1840*, 266; see a similar view from Lord Farnborough in Carlisle, Cumbria Record Office, D/LONS/L1/2/50.

39. *KW* vi. 573–5. The Ladies' Gallery was added in a second wave of repairs in July 1835.

40. *Journal of the House of Lords*, lxvi (London, 1834), 997–1000. Parliament was prorogued again on 25 November and 18 December; the venue of the ceremony is not known, but possibly once again the Lords Library.

41. *KW* vi. 574.

42. *Returns of the Expense of Erecting and Furnishing the present Temporary Houses of Parliament*, House of Commons, 1836 (499), 2–3; see also *KW* vi. 666.

43. *KW* vi. 575; M. H. Port, *The Houses of Parliament* (London, 1976), 146.
44. *Parliamentary Debates*, 3rd Series, xxvi, cols. 1–3.
45. Ibid., cols. 63–4.
46. Sainty, 'The Office of Housekeeper of the House of Lords', 258–9.
47. *KW* vi. 680–1.
48. *KW* vi. 223.
49. Reynolds, 'FitzClarence, George', *ODNB*.
50. Given on 27 June 1855 at the Drury Lane Theatre: *The Speeches of Charles Dickens*, ed. K. J. Fielding (Oxford, 1960), 204–5.
51. Michael Goldberg, 'From Bentham to Carlyle: Dickens' Political Development', *Journal of the History of Ideas*, 33 (1972), 64.
52. Yelling, *Slums*, 20, 26; see www.peabody.org.uk.
53. *The Examiner*, 19 October 1834.
54. *Report…on Record Commission*, 33, 436–7.
55. Ibid. 100.
56. *Report of the Commissioners appointed to inquire into the state of the Public Records*, House of Commons Sessional Paper, 1837 (71), 8.
57. *Public Record Office Act*, 1 & 2 Victoria (1838), *c.* 94.
58. Hallam, 'Nine Centuries of Keeping the Public Records', 38; Martin, 'Palgrave, Sir Francis (1788–1861)', *ODNB*.
59. Bernard Nurse, 'Black, William Henry (1808–1872)', *ODNB*.
60. Ann Cooper, 'Cole, Sir Henry (1808–1882)', *ODNB*.
61. *Gentleman's Magazine*, December 1834, p. 630.
62. Anita McConnell, 'Simms, William (1793–1860), *ODNB*.
63. *Report of the Commissioners…of the standards of weight & measure*, 10–13.
64. A. J. Taylor, *The Jewel Tower, Westminster* (London, 1965), 17–18.
65. Williams, *Life and Letters of John Rickman*, 316–19.
66. *KW* vi. 534.
67. *KW* vi. 525, 534, 625.
68. *KW* vi. 624.
69. *KW* vi. 247 ff.
70. Port, *Houses of Parliament*, 178–80.
71. *Report of the…Fire*, 51.
72. *KW* vi. 220 n. 3, 307, 326, 393, 398, 422, 478, 561, 571.
73. *Morning Chronicle*, 8 November 1834.
74. *The Examiner*, 19 October 1834.
75. *Morning Chronicle*, 18 October 1834; *Caledonian Mercury*, 20 October 1834; *Ipswich Journal*, 18 October 1834.

76. *New Monthly Magazine,* November 1834, p. 356.
77. *KW* vi. 573, 575.
78. *Gentleman's Magazine,* November 1834, p. 480; *The Times,* 29 October 1834, p. 3, Letter from 'Economy'.
79. Port, *Houses of Parliament,* 146.

BIBLIOGRAPHY

Manuscripts

Bedfordshire and Luton Archives and Record Service, Bedford: Z629/41.
British Library, London: Egerton MS 1084; Fox-Talbot Papers.
Centre for Buckinghamshire Studies, Aylesbury: D/FR/119/13/1/9.
Centre for Kentish Studies, Maidstone: U840 C534/2.
City of Westminster Archives Centre, London: Box 57; Box E133; Greenwood map of London, 1827; Horwood/Faden map of London, 1813.
Cumbria Record Office, Carlisle: D/LONS/L1/2/50.
Durham University Library, Durham: GRE/B41/2/145/2.
Guildhall Library, London: MS 15728/2.
Herefordshire Record Office: G2/IV/J/66.
London Metropolitan Archives, London: LME/01 no. 498.
Museum of London, London: Accession A15453.
National Archives, London (TNA): E 402; HO 44/27, 29; PC 1/2515; WORK 11/26/5; WORK 29/20–7, 78, 763–4, 766, 769–70.
Norfolk Record Office, Norwich: MC 81/26/222 525x8.
Parliamentary Archives, London (PA): HC/SO/1/6; HC/SA/SJ/9/52, 55; HL/PO/1/1; HL/PO/LB/2/1; HL/PO/RO/1/78, 90, 158, 195; OOW/3; RIC/1-3; LGC/5/3, 10/1/1.
Trinity College, Cambridge: O.14.13/105, 112.

Broadsheets

A Conversation Between the Abbey and Westminster Hall, Printed by T. Birt, at St Andrew St, Seven Dials, London [1834] (University of Chicago Library, Illinois, USA, Song and Broadsheet Collection, box 1, fo. 9).

Dreadful Fire! And total destruction of both Houses of Parliament, G. Drake [1834] (Westminster City Archives, E133 (106)).

Fatal destruction of both Houses of Parliament, Thursday evening, October 16th 1834, Carpue Printer 3 Old Montague Street [1834] (British Library, C.116.i.2, fo. 29).

The Rambler's Flash Songster, nothing but out and outers, adapted for gentlemen only, etc. (London: [1838?]).

View of the Destruction of both Houses of Parliament, by fire; and immense loss of national property from a drawing taken on the spot by an eminent artist, London, published by J. Fairbairn, 110 Minories [1834] (Parliamentary Works of Art Collection, WOA/2978a).

Newspapers And Periodicals

The Bristol Mercury

Caledonian Mercury

Children's Daily Telegraph

Cobbett's Weekly Register

The Examiner

The Gentleman's Magazine

Ipswich Journal

Manchester Times & Gazette

Metropolitan Magazine

The Morning Chronicle

The Morning Post

New Monthly Magazine

New Sporting Magazine

Preston Chronicle

The Spectator

The Standard

The Times

Other Printed Works

The A to Z of Regency London, ed. Paul Laxton and Joseph Wisdom (London, 1985).

Allen, Thomas, *The History and Antiquities of London, Westminster and Southwark and parts adjacent*, 5 vols. (London, 1828).

Ascoli, D., *The Queen's Peace: The Origins and Development of the Metropolitan Police* (London, 1979).

Balston, T., *The Life of Jonathan Martin, Incendiary of York Minster* (London, 1945).

Barry, Alfred, *Memoir of the life and works of the late Sir Charles Barry, Architect*, 2nd edn (London, 1870).

Bernal, Ralph, 'My Aunt Mansfield', *The Keepsake* (London, 1835).

Binski, P., *The Painted Chamber at Westminster*, Society of Antiquaries Occasional Paper (New Series), ix (1986).

Bibliography

Bonython, Elizabeth, and Burton, Anthony, *The Great Exhibitor: The Life and Work of Henry Cole* (London, 2003).

Bradley, Simon, and Pevsner, Nicholas, *The Buildings of England. London 6: Westminster* (New Haven and London, 2003).

Braidwood, James, *On the construction of fire-engines and apparatus, the training of firemen and the method of proceeding in cases of fire* (Edinburgh, 1830).

——*Fire Prevention and Fire Extinction* (London, 1866).

Brayley, E. W., and Britton, J., *The History of the Ancient Palace and Late Houses of Parliament at Westminster* (London, 1836).

Brock, Michael, *The Great Reform Act* (London, 1973).

Broughton, Lord, *see* Hobhouse, J. C.

Bush, M. L., *The Casualties of Peterloo* (Lancaster, 2005).

Campbell, John, Lord, *Life of John, Lord Campbell*, edited by his daughter Mrs Hardcastle, 2 vols. (London, 1881).

Cantwell, J. D, *The Public Record Office, 1838–1958* (London, 1991).

Carlyle, Thomas, *Letters of Thomas Carlyle 1826–1836*, ed. Charles Eliot Norton (London, 1889).

Clanchy, M. T., *From Memory to Written Record: England 1066–1307*, 2nd edn (Oxford, 1993).

Cobb, H. S., 'The Victoria Tower and the Records of Parliament, 1864–1986', *The Table: The Journal of the Society of Clerks at the Table in Commonwealth Parliaments*, liv (1986), 54–63.

Colvin, H. M., 'Views of the Old Palace of Westminster', *Architectural History*, 9 (1966), 21–184.

——et al., *The History of the King's Works*, 6 vols. (London, 1963–82). [Abbreviated as *KW*]

Cooke, Robert, *The Palace of Westminster: Houses of Parliament*, (London, 1987).

Cormack, Patrick, *Westminster: Palace and Parliament* (London, 1981).

——'The Great Fire of Westminster 1834', *The Historian* (Autumn, 1984), 3–6.

Crewe, Sir George, *Sir George Crewe, Squire of Calke Abbey: Extracts from the Journals of Sir George Crewe, 1815–1834*, ed. Colin Kitching (Matlock, 1995).

Crook, John, and Harris, Roland B., 'Reconstructing the Lesser Hall: An Interim Report from the Medieval Palace of Westminster Research Project', in Clyve Jones and Sean Kelsey (eds), *Housing Parliament*, Parliamentary History Yearbook Trust (Edinburgh, 2002), 22–61.

Cunningham, Peter, *Murray's Handbook for Modern London, or London as it is* (London, 1851).

Darley, Gillian, *John Soane: An Accidental Romantic* (London, 1999).

Dasent, Arthur Irwin, *The Speakers of the House of Commons* (London and New York, 1911).

Dialogus de Scaccario, ed. C. Johnson (Oxford, 1983).

Dickens, Charles, *Sketches by Boz* (London, 1836).

—— *Oliver Twist* (London, 1837).

—— *The Letters of Charles Dickens*, i (1820–1839), ed. Madeline House and Graham Storey (Oxford, 1965).

—— *Dickens' Journalism*, ed. Michael Slater, 4 vols. (London, 1996–2000).

Earle, C. W. [Maria Theresa], *Memoirs and Memories: Pot-pourri from a Surrey Garden* (London, 1911).

Escott, Margaret, 'The Procedure and Business of the House', in D. R. Fisher (ed.), *The History of Parliament: The House of Commons 1820–1832*, 7 vols. (Cambridge, 2009), i. 283–318.

Farrell, Stephen, 'The Armada Tapestries in the Old Palace of Westminster', *Parliamentary History*, 29 (2010), 416–40.

Ferrey, Benjamin, *Recollections of A. W. N. Pugin and his Father Augustus Pugin*, (London, 1861; Scolar Press reprint, 1978).

Field, John, *The Story of Parliament in the Palace of Westminster* (London, 2002).

First Report of the Royal Commission on Historical Manuscripts, Command Paper C.55 (London, 1874).

Flora Tristan's London Journal: A Survey of London Life in the 1830s. A Translation of Promenades dans Londres, ed. Dennis Palmer and Giselle Pincetl (London, 1980).

Fredericksen, Andrea, 'Parliament's Genius Loci: The Politics of Place after the 1834 Fire', in C. Riding and J. Riding (eds), *The Houses of Parliament: History, Art, Architecture* (London, 2000), 99–111.

Galinou, Mireille, 'Adam Lee's Drawings of St Stephen's Chapel, Westminster: Antiquarianism and Showmanship in Early 19th Century London', *Transactions of the London and Middlesex Archaeological Society*, 34 (1983), 231–44.

Gerhold, Dorian, *Westminster Hall: Nine Hundred Years of History* (London, 1999).

Goldberg, M., 'From Bentham to Carlyle: Dickens' Political Development', *Journal of the History of Ideas*, 33/1 (1972), 61–76.

Goodall, John, 'The Medieval Palace of Westminster', in C. Riding and J. Riding (eds), *The Houses of Parliament: History, Art, Architecture* (London, 2000), 49–67.

Gosse, Edmund, *Catalogue of the Library of the House of Lords* (London, 1908).

Grant, James, *Random Recollections of the House of Lords from the year 1830 to the close of 1836* (London, 1836).

——*Random Recollections of the House of Commons from the year 1830 to the close of 1835*, 5th edn (London, 1837).

Greenhead, John, 'History of the House of Lords Library', *House of Lords Library Note LLN 2009/005* (London, 2009).

The Greville Memoirs 1814–1860, ed. L. Strachey and R. Fulford, 8 vols. (London, 1938).

Guide to the Contents of the Public Record Office, 3 vols. (HMSO, London, 1963).

Hahn, Daniel, *The Tower Menagerie* (London, 2003).

Hallam, Elizabeth M., 'Nine Centuries of Keeping the Public Records', in G. H. Martin and P. Spufford (eds), *The Records of the Nation* (Woodbridge, 1990), 23–42.

——and Roper, M., 'The Capital and the Records of the Nation: Seven Centuries of Housing the Public Records in London', *The London Journal*, 4 (1978), 73–94.

Hallam Smith, Elizabeth, 'The Chapter House as a Record Office', in Warwick Rodwell and Richard Mortimer (eds), *Westminster Abbey Chapter House: The History, Art and Architecture of 'a chapter house beyond compare'* (London, 2010), 124–38.

Hamilton, David C., *The Assassination of the Prime Minister: John Bellingham and the Murder of Spencer Perceval* (Stroud, 2008).

Harris, A. T., *Policing the City, Crime and Legal Authority in London, 1780–1840* (Columbus, Ohio, 2004).

Hastings, Maurice, *Parliament House: The Chambers of the House of Commons* (London 1950).

Hawkyard, Alasdair, 'From Painted Chamber to St Stephen's Chapel: The Meeting Places of the House of Commons at Westminster until 1603', in Clyve Jones and Sean Kelsey (eds), *Housing Parliament*, Parliamentary History Yearbook Trust (Edinburgh, 2002), 62–84.

Hay, Malcolm, 'The Armada Paintings in the New Palace of Westminster', *Parliamentary History*, 29 (2010), 441–51.

——and Kennedy, Rachel, *The Medieval Kings c1388. An Exhibition of Three Statues from the South Wall. Their History and Recent Conservation*, exhibition catalogue (London, 1993).

Haydon, Benjamin, *Life of Benjamin Robert Haydon, Historical Painter, from his Autobiography and Journals*, ed. Tom Taylor, 3 vols. (London, 1853).

Henham, Brian, *True Hero: The Life and Times of James Braidwood, Father of the British Fire Service* (Romford, 2000).

Hickin, W. F., *Organised against Fire: A Short Organisational History of the London Fire Brigade and its Predecessors from 1833 to 1996* (London, 1996).

311

Hill, Rosemary, *God's Architect: Pugin and the Building of Romantic Britain* (London, 2008).

Hobhouse, J. C., *Recollections of a long life by Lord Broughton (John Cam Hobhouse) with additional extracts from his private diaries*, ed. Lady Dorchester, 6 vols. (London, 1911).

Hobsbawm E., and Rudé, G., *Captain Swing* (London, 1969).

The Holland House Diaries, 1831–1840, ed. Abraham D. Kriegel (London, 1977).

Holland, Michael (ed.), *Swing Unmasked* (Milton Keynes, 2005).

Holloway, Sally, *Courage High! A History of Firefighting in London* (London, 1992).

'The House of Commons refreshment department', House of Commons Information Office, Factsheet G19, General Series (London, 2003).

Hughson, David, *Walks through London including Westminster and the borough of Southwark*, 2 vols. (London, 1817).

Jay, Winifred, 'The House of Commons and St. Stephen's Chapel', *English Historical Review*, xxxvi (1921), 225–8.

Jenkinson, Hilary, 'Exchequer Tallies', *Archaeologia*, 62 (1911), 369–70.

—— 'Medieval Tallies, Public and Private', *Archaeologia*, 74 (1925), 329–51.

Johnson, D. J., 'The House of Lords and its Records, 1660–1864', in H. S. Cobb (ed.), *Parliamentary History, Libraries and Records: Essays Presented to Maurice Bond* (House of Lords Record Office, London, 1981), 25–32.

Jones, Christopher, *The Great Palace* (London, 1983).

Jones, D. L., 'The House of Lords and the Chambre des Pairs: An Exchange of Books, 1834', in H. S. Cobb (ed.), *Parliamentary History, Libraries and Records: Essays Presented to Maurice Bond* (House of Lords Record Office, London, 1981), 51–8.

Lee, Adam, *Description of the Cosmoramic and Dioramic Delineations of the Ancient Palace of Westminster and St Stephen's Chapel* (London, 1831).

Lees, Lynn Hollis, *Exiles of Erin: Irish Migrants in Victorian London* (Manchester, 1979).

Leslie, C. R., *Memoirs 00066 the Life of John Constable*, ed. J. Mayne, 3rd edn (London, 1995).

Liversidge M., and Binski, P., 'Two Ceiling Fragments from the Painted Chamber at Westminster Palace' *The Burlington Magazine*, 137 (1995), 491–501.

Low, Donald A., *The Regency Underworld* (Stroud, 2005).

McKay, W. R., *Clerks in the House of Commons 1363–1989: A Biographical List*, House of Lords Record Office Occasional Publications No. 3 (HMSO, London, 1989).

Marshall, P. H., *Charles Marshall RA, his Origins, Life and Career* (privately printed, Truro, 2000).

Matthews, Mike, *Captain Swing in Sussex and Kent: Rural Rebellion in 1830* (Hastings, 2006).

Menhennet, D., *The House of Commons Library: A History* (London, 1991).

Minutes of evidence taken before the select committee on the bill to amend and render more effectual two acts of the fifth and sixth years of the reign of his late Majesty King George the Fourth, relating to Weights and Measures, House of Commons Sessional Papers, 1834 (464).

Mitchell, L. G., 'Foxite Politics and the Great Reform Act', *English Historical Review*, 108 (1993), 338–64.

Munby, A. N. L., *Phillipps Studies*, 2 vols. (London, 1971).

Ormrod, W. M., *The Reign of Edward III: Crown and Political Society in England 1327–1377* (New Haven and London, 1990).

Orsinger, T. J., and Orsinger, D. F., *The Firefighter's Best Friend: Lives and Legends of Chicago Firehouse Dogs* (Chicago, 2003).

Oxford Dictionary of National Biography (Oxford, 2004 [abbreviated as *ODNB*], and online edition).

Palmer, Stanley H., *Police and Protest in England and Ireland 1780–1850* (Cambridge, 1988).

Parliamentary Debates, 3rd series (London, 1832–4).

Pearce, Edward, *Reform! The Fight for the 1832 Reform Act* (London, 2003).

Peel, Sir Robert, *Sir Robert Peel from his Private Papers*, ed. C. S. Parker, 3 vols. (London, 1899).

Port, M. H., *The Houses of Parliament* (London, 1976).

Porter, Roy, *London: A Social History* (London, 2000).

Pugin, A. W. N., *The Collected Letters of A. W. N. Pugin*, ed. Margaret Belcher, 2 vols. (Oxford, 2001).

Raikes, T., *A Portion of the Journal Kept by Thomas Raikes Esq, from 1831 to 1847*, 4 vols. (London, 1856).

Read, D., *Peterloo: The Massacre and its Background* (Manchester, 1958).

Report from the Select Committee on House of Commons buildings, together with the minutes of evidence taken before them, House of Commons Sessional Papers, 1831 (308).

Report from the Select Committee on the House of Commons buildings with the minutes of evidence taken before them, House of Commons Sessional Papers, 1833 (269).

Report from the Select Committee on the Losses of the Late Speaker and Officers of the House by Fire of The Houses of Parliament; with the minutes of evidence and appendix, House of Commons Sessional Papers, 1837 (493).

Report from the Select Committee on Plans for the Permanent Accommodation of Parliament, House of Lords Sessional Papers, 1835 (73).

Report from the Select Committee on Record Commission, House of Commons Sessional Papers, 1836 (429) (565).

Report of the Commissioners appointed to consider the steps to be taken for restoration of the standards of weight & measures, House of Commons Sessional Papers, 1842 (356).

Report of the Commissioners appointed to inquire into the state of the Public Records, House of Commons Sessional Paper, 1837 (71).

Report of the Lords of the Council respecting the Destruction by Fire of the Two Houses of Parliament: with the minutes of evidence, House of Commons Sessional Papers, 1835 (1). [Abbreviated as *Report of the...Fire*]

Reports from the Select Committee appointed to inquire into the state of the public records of the Kingdom etc (London, 1800).

Returns of the Expense of Erecting and Furnishing the present Temporary Houses of Parliament, House of Commons Sessional Papers, 1836 (499).

Rogers, Phillis M., 'Medieval Fragments from the Old Palace of Westminster in the Sir John Soane Museum', in H. S. Cobb (ed.), *Parliamentary History, Libraries and Records: Essays Presented to Maurice Bond* (House of Lords Record Office, London, 1981), 1–8.

Sainty, J. C., *Officers of the Exchequer*, List and Index Society, Special Series, 18 (London, 1983).

—— *The Parliament Office in the Nineteenth and Early Twentieth Centuries: Biographical Notes on Clerks in the House of Lords 1800 to 1939* (House of Lords Record Office, London, 1990).

—— 'The Office of Housekeeper of the House of Lords', *Parliamentary History*, 27 (2008), 256–60.

Salmon, Philip, *Electoral Reform at Work: Local Politics and National Parties 1832–1841* (Woodbridge, 2002).

—— 'The English Reform Legislation, 1831–1832', in D. R. Fisher (ed.), *The History of Parliament: The House of Commons 1820–1832*, 7 vols. (Cambridge, 2009), i. 374–412.

Sebba, Anne, *The Exiled Collector: William Bankes and the Making of an English Country House* (Wimborne Minster, 2009).

[Shenton, Caroline], 'Introductory Note: Exchequer of Receipt: Original Receipts: E 402', (National Archives, London, 1995).

Shenton, Caroline, Prior, David, and Takayanagi, Mari, *Victoria Tower Treasures from the Parliamentary Archives* (London, 2010).

Smirke, Sydney, 'Remarks on the Architectural History of Westminster Hall', *Archaeologia*, 26 (1836), 406–21.

Soane, John, *Designs for public buildings* (London, 1828).

——*Memoirs of the professional life of an architect between the years 1768 and 1835 written by himself* (London, 1835).

Solender, K., *Dreadful Fire! Burning of the Houses of Parliament* (Cleveland Museum of Art, Cleveland, Ohio, 1984).

Stanley, A. P., *Historical Memorials of Westminster Abbey*, 2nd edn (London, 1868).

Strong, Roy, *Lost Treasures of Britain* (London, 1990).

Taylor, A. J., *The Jewel Tower, Westminster* (London, 1965).

Thomas, Christopher, Cowie, Robert, and Sidell, Jane, *The Royal Palace, Abbey and Town of Westminster on Thorney Island: Archaeological Excavations (1991–8) for the London Underground Ltd Jubilee Line Extension Project*, Museum of London Archaeological Service Monograph 22 (London, 2006).

Thompson, Dorothy, *The Chartists: Popular Politics in the Industrial Revolution* (Aldershot, 1984).

Tinniswood, Adrian, *The Polite Tourist: A History of Country House Visiting* (London, 1998).

Tite, Colin G. C., 'The Cotton Library in the Seventeenth Century and its Manuscript Records of the English Parliament', *Parliamentary History*, xiv (1995), 121–8.

Tomalin, Claire, *Mrs Jordan's Profession: The Story of a Great Actress and a Future King* (London, 2003).

Torrens, W. M., *Memoirs of William Lamb, Second Viscount Melbourne*, new rev. edn (London, 1890).

Walker, J. R. B., 'The Palace of Westminster after the fire of 1834', *Annual Volume of the Walpole Society*, 44 (1975), 94–122.

Walne, Peter, 'The Record Commissions, 1800–1837', *Journal of the Society of Archivists* (1960), 8–16.

Walthew, Kenneth, *From Rock and Tempest: The Life of Captain George William Manby* (London, 1971).

——'Captain Manby and the Conflagration of the Palace of Westminster', *History Today*, 32 (April 1982), 21–5.

Weitzman, George H., 'The Utilitarians and the Houses of Parliament', *The Journal of the Society of Architectural Historians*, 20 (1961), 99–107.

Werrett, Simon, *Fireworks: Pyrotechnic Arts and Sciences in European History* (Chicago and London, 2010).

White, Jerry, *London in the 19th Century* (London, 2007).

Wilkinson, Clare, 'Politics and Topography in the Old House of Commons, 1783–1834', in Clyve Jones and Sean Kelsey (eds), *Housing Parliament*, Parliamentary History Yearbook Trust (Edinburgh, 2002).

Williams, O. C., *The Officials of the House of Commons* (London, 1909).

——*Lamb's Friend the Census Taker: Life and Letters of John Rickman* (London, 1912).

——*The Topography of the Old House of Commons 1769–1774* (unpublished monograph, House of Lords Record Office, 1953).

——*The Clerical Organisation of the House of Commons, 1661–1850* (Oxford, 1954).

Wilson, A. N., *The Victorians* (London, 2002).

Wright, Brian, *A Fire in the House: A Dramatic Account of the Fire which Destroyed the Old Houses of Parliament in October 1834* (privately printed, n.p., 1986).

Yelling, J. A., *Slums and Slum Clearance in Victorian London* (London, 1986).

Ziegler, Philip, *William IV* (London, 1971).

INDEX

Index